Mordecai Cubitt Cooke, M. C. (Mordecai Cubitt) Cooke

Freaks and marvels of plant life

or, Curiosities of vegetation

Mordecai Cubitt Cooke, M. C. (Mordecai Cubitt) Cooke

Freaks and marvels of plant life
or, Curiosities of vegetation

ISBN/EAN: 9783337375041

Printed in Europe, USA, Canada, Australia, Japan

Cover: Foto ©berggeist007 / pixelio.de

More available books at **www.hansebooks.com**

FREAKS AND MARVELS

OF

PLANT LIFE;

OR,

CURIOSITIES OF VEGETATION.

BY

M. C. COOKE, M.A. LL.D.
AUTHOR OF
"PONDS AND DITCHES," "THE WOODLANDS," ETC. ETC.

FOURTH THOUSAND.

PUBLISHED UNDER THE DIRECTION OF
THE COMMITTEE OF GENERAL LITERATURE AND EDUCATION
APPOINTED BY THE SOCIETY FOR PROMOTING
CHRISTIAN KNOWLEDGE.

LONDON
SOCIETY FOR PROMOTING CHRISTIAN KNOWLEDGE,
NORTHUMBERLAND AVENUE, CHARING CROSS, S.W.;
43, QUEEN VICTORIA STREET, E.C.; 48, PICCADILLY, W.;
AND 135, NORTH STREET, BRIGHTON.
NEW YORK: E. & J. B. YOUNG & Co.
1882.

CONTENTS.

CHAPTER		
I.	INTRODUCTION	page 1
II.	THE SUNDEWS	23
III.	VENUS'S FLY-TRAP	50
IV.	SIDE-SADDLE FLOWERS	72
V.	PITCHER-PLANTS	95
VI.	MINOR CARNIVORA	122
VII.	GYRATION OF PLANTS	150
VIII.	HELIOTROPES, OR SUNFLOWERS	170
IX.	TWINERS AND CLIMBERS	184
X.	SENSITIVE PLANTS	220
XI.	SLEEP OF PLANTS	239
XII.	METEORIC FLOWERS	259
XIII.	HYGROSCOPISM	275
XIV.	DISPERSION	291
XV.	MIMICRY	321
XVI.	GIANTS	345
XVII.	TEMPERATURE	371
XVIII.	LUMINOSITY	383
XIX.	MYSTIC PLANTS	401
XX.	FLOWERS OF HISTORY	446

LIST OF ILLUSTRATIONS.

FIG.		PAGE
1.	Round-leaved Sundew, *Drosera rotundifolia*	24
2.	Section of gland of *Drosera rotundifolia*, magnified	25
3.	Leaf with tentacles of *Drosera rotundifolia*, enlarged	30
4.	Venus's Fly-trap, *Dionæa muscipula*	51
5.	Leaf of *Aldrovanda*, enlarged	68
6.	Glands on leaf of *Drosophyllum*, magnified	69
7.	Pitchers of *Sarracenia variolaris*, reduced	76
8.	Pitcher of *Sarracenia purpurea*, reduced, with section	77
9.	*Sarracenia purpurea*	78
10.	Pitchers of *Darlingtonia*	96
11.	Pitcher of *Nepenthes bicalcarata*	102
12.	Pitcher of *Nepenthes Chelsoni*	110
13.	Section of hood of *Nepenthes Chelsoni*	111
14.	Pitcher of *Cephalotus*	115
15.	Section of pitcher of *Cephalotus*	116
16.	Glands of *Cephalotus*, in section	117
17.	Butterwort (*Pinguicula Lusitanica*)	123
18.	Leaf of Butterwort with the edges curved inwards	126
19.	Bladderwort (*Utricularia vulgaris*)	131
20.	Bladder of *Utricularia vulgaris*, enlarged	132
21.	Trifolium subterraneum fruit	166
22.	Natal Climbing Plant (*Ceropegia Sandersoni*)	192
23.	Bitter-Sweet (*Solanum dulcamara*)	193
24.	The Twining Polygonum (*Polygonum convolvulus*)	195
25.	Leaf of *Bomarea Carderi*, the petiole twisted in the reversal of the leaf	197
26.	Traveller's Joy (*Clematis vitalba*)	200
27.	Swollen petiole of *Clematis vitalba*	202
28.	Common Fumitory (*Fumaria officinalis*)	204
29.	Climbing Corydalis (*Corydalis claviculata*)	205

LIST OF ILLUSTRATIONS.

FIG.		PAGE
30.	Hooked tendril, like foot of a bird, from *Bignonia Tweediana*. Tip of hook magnified	210
31.	Tendrils of Virginia creeper, with discs attached	212
32.	Tendrils of Virginia creeper, discs not attached	212
33.	Tendrils of *Passiflora edulis*	214
34.	Cleavers (*Galium aparine*)	215
35.	Leaves of sensitive plant, *Mimosa pudica*, awake and asleep	222
36.	Grass of Parnassus (*Parnassia palustris*)	233
37.	Flowers of Epilobium	234
38.	Leaves of Wood-sorrel	244
39.	Leaflets of Clover, awake and asleep	247
40.	Leaf of *Acacia Farnesiana*, awake	253
41.	Leaf of *Acacia Farnesiana*, in a sleeping condition	253
42.	Scarlet Pimpernel (*Anagallis arvensis*)	263
43.	Evening Primrose (*Œnothera biennis*)	266
44.	Bee Orchis (*Ophrys apifera*)	269
45.	Snipe Orchis	270
46.	Flowers of *Pachystoma Thomsoni*	270
47.	*Dendrobium D'Albertisii*	271
48.	Zebra Orchis (*Oncidium zebrinum*)	272
49.	Wild oat (*Avena fatua*)	276
50.	Capsules of *Mesembryanthemum tripolium* closed	283
51.	Capsule of *Mesembryanthemum tripolium* open	283
52.	Sand-box (*Hura crepitans*)	284
53.	Balsam (*Impatiens*)	293
54.	Caltrops, or fruits of *Tribulus terrestris*	299
55.	Fruits of *Pedalium murex*	299
56.	Burdock (*Lappa minor*)	300
57.	Hooked fruits of *Martynia diandra*	302
58.	Fruit of *Proboscidea Jussieui*, reduced	303
59.	Fruit of Grapnel plant, natural size (*Harpagophytum leptocarpum*)	304
60.	Fruit of *Trapa bicornis*	305
61.	Fruit of *Trapa bispinosa*	305
62.	Fruit of *Gahnia xanthophylla*, suspended	307
63.	Receptacle of the Egyptian Bean (*Nelumbium speciosum*)	308
64.	Monkey Pots (*Lecythes sp.*)	309

FIG.		PAGE
65.	Cannon Ball (*Couroupita guianensis*)	314
66.	Euphorbia, resembling a Cactus growing amongst rocks in Damara Land	325
67.	Young plants of *Salvinia, Jussiæa repens, Phyllanthus*	329
68.	*Lycopodium compactum*	327
69.	*Azorella selago*	327
70.	Leaf of *Caltha dionæfolia*	328
71.	*Actinotus*	329
72.	Rock Rose (*Helianthemum*)	330
73.	Seeds of *Messua ferrea*, natural size	331
74.	Samara of *Securidaca tomentosa, Heteropterys laurifolia, Gallesia gorancma, Seguiera floribunda*	332
75.	Samara of *Ulmus campestris, Ulmus montana, Ptelea trifoliata, Hiræa*	333
76.	Seed of *Calosanthes indica*	334
77.	Seed of *Zanonia macrocarpa*	334
78.	Crested seed of *Sarcostemma, Echites scabra*, Willow-herb (*Epilobium*), Milk Thistle (*Silybum marianum*)	336
79.	Snake nut (*Ophiocaryon serpentintium*) cut open	343
80.	Giant Arum (*Amorphophallus Titanum*) greatly reduced	359
81.	*Rafflesia Arnoldi*, from a photograph of the living flower	360
82.	Flower of *Aristolochia Goldieana* reduced	363
83.	Wake-robin (*Arum maculatum*)	375
84.	Egyptian Lotus (*Nymphæa stellata*)	409
85.	Lady with Lotus Flower, from Theban tomb	411
86.	Daffodil (*Narcissus pseudonarcissus*)	415
87	Jesuitic Maracoc	432
88.	Passion Flower (*Passiflora cincinnata*)	436
89.	*Medicago echinus*	437
90.	Mistletoe (*Viscum album*)	438
91.	Male Mandrake	440
92.	Female Mandrake	442
93.	Broom (*Sarothamnus scoparius*)	448
94.	Cotton Thistle (*Onopordum acanthium*)	453
95.	Musk Thistle (*Carduus nutans*)	454
96.	Scotch coin of 1602	455
97.	Scotch coin of 1599	455

FREAKS OF PLANT LIFE.

CHAPTER I.

INTRODUCTION.

THIS work has been undertaken for the purpose of presenting in a popular form, devoid as much as possible of technical language, some of the most prominent features in the investigations which have of late years contributed so much to our knowledge of the phenomena of vegetable life. The labours especially of Mr. Darwin in this direction deserve to be more generally known than they are. Unfortunately, the dread which non-scientific persons exhibit at the outside of a scientific book often prevents any attempt at understanding its contents. Hence we have made an effort to summarise the results of these and similar experiments, and to present in as succinct a manner as the subjects permitted, their teachings. Some elaborate investigations, as, for instance, those on fertilisation, are chiefly of interest to botanists, and could be little understood or appreciated by the general public;

these have, therefore, not been considered as falling within the limits of this volume. On the other hand, chapters are introduced on subjects which have not yet been submitted to exhaustive examination, but which have, nevertheless, great popular interest and fall legitimately within the scope of the title. Free use has been made of all sources of information, under the conviction that the better these experiments are known and understood, the greater and more general will be the appreciation of the labours of those who have contributed so much to the elucidation of obscure phenomena in plant-life.

Text-books remind us of the importance of the vegetable world in its relationship to the animal. They also illustrate the grandeur and beauty which the plant has conferred on the world. It is difficult to form any adequate conception of the vast extent and unlimited variety of vegetable life. All we can do is to pick up here and there some object of special interest, gaze at it, marvel at it, try to comprehend it, if we can, and then pass on, leaving behind us a trackless ocean of wonderful things, to be picked up by our successors, and marvelled at as we have done. It will be very long before the storehouse is exhausted.

We learn to appreciate what has been written of wild forests only by experience. "A very similar feeling (to that of a sea-voyage) possesses the traveller as he penetrates an extensive forest. Every morning he commences his journey, patiently pursuing the winding pathways through interminable multitudes of trees and shrubs, till, when evening arrives, he is hardly less fatigued with the monotony of the scene than with the exertions of the day. His feelings are the same as those at sea—he is surprised at the interminable character of the scene, and his ideas of space are measured by a greater standard. He wonders at the vast multitudes of vegetable beings; whence they could possibly have drawn nourishment to rear such solid structures; he speculates on their age, and lastly on their use. In both cases the ideas of space are the same, but they have received an impulse from the novelty of the scene; perhaps assisted also by the perfect stillness, which reigns so completely in deep forests, and during the heat of the day the silence is more painful than on the wide ocean. The chief difference between the two is that one is a sea of waters, the other a sea of trees."[1]

It is a very natural inquiry, and one which may be fairly considered as a prelude to a subject such as ours, what number of different kinds, or species, of plants are supposed to be found on the surface of the globe? This is a question which has been pro-

---

[1] Hinds in "Annals of Nat. Hist.," xv. (1845), p. 89.

pounded before, and more than once its solution has been attempted.[1] The history of these progressive estimates it rather a curious one. It commences 390 B.C. with Theophrastus, and he enumerated 500 kinds of plants. This may be presumed to represent all that were then known. The botanical knowledge of King Solomon had, then, comparatively narrow limits, even though he discoursed on all plants from the cedar of Lebanon to the hyssop on the wall. Pliny (A.D. 79) increased the number of plants to double that of his predecessor. In the beginning of the seventeenth century the number had increased to 6,000. The second edition of Linnæus's great book included no more than 8,800. Willdenow, up to 1807, had detected 17,457 species of flowering plants. From this period the increase in the number of known species was very rapid, as a result of the stimulus given to botany by Linnæus and his successors, so that at the beginning of the present century Robert Brown had calculated the flowering plants at 37,000, and Humboldt all plants, flowering and non-flowering, at 44,000.

Progressing still further down the stream of time,

---

[1] R. B. Hinds on "Geographical Botany," "Annals of Nat. Hist.," xv. (1845), p. 15. A. Henfrey, "Elementary Course of Botany" (1857), p. 659. Humboldt, "Views of Nature" (1850), p. 276.

in 1820, De Candolle calculated that at the least 56,000 species of plants were known. It was found that the number of species preserved in the Herbarium at the Jardin des Plantes was estimated at the same figure, and that the collection of M. Delessert contained as many as 86,000 species in 1847, although Dr. Lindley had estimated in 1835 that all the plants in the world might be included in that number.

Humboldt entered upon a series of calculations, about this time, to show that all these estimates fell short of the number which might be supposed to exist. "Such considerations," he writes, "which I purpose developing more fully at the close of this illustration, seem to verify the ancient myth of the Zend-Avesta that the creating primeval force called forth 120,000 vegetable forms from the sacred blood of the bull." In 1845 Mr. R. B. Hinds estimated the total of phanerogamic and cryptogamic plants at 134,000 species. The next estimate we meet with is in Henfrey (1857) at 213,000, but in 1855 De Candolle had, by another process of reasoning, come to the conclusion that the total could not be less than 375,000 for flowering plants. Doubtless, these calculations will go on increasing, as the highest is found to be inadequate to represent even the whole number of described species. At the present time the very lowest estimate of authentic species of cryptogamia cannot be less than 50,000, and probably considerably exceed that number.

Here, then, we have somewhat of an approximate idea, at what may be regarded a very low estimate, of the number of species of plants scattered over the face of the earth. It is always best in such calculations to under-estimate rather than over-estimate, and if we feel confidence in asserting that there are not less than 500,000 distinct and different species of vegetable organisms distributed over the globe, including land and water, it is because we feel satisfied that we have good grounds for believing that the number is in excess even of that which we have permitted ourselves to affirm.

Another "curiosity" is in respect to the relative sizes of plants; some we know to be very large, and others are very small, what then is the average size? It has been calculated, in the animal world, that between the largest living animal known on the one hand, and the smallest which the microscope has revealed, the middle place, between both extremes, is occupied by the common house-fly. If we pursue a similar plan with plants, and estimate the smallest flowering plant to be the little Duckweed (*Lemna minor*), and the largest a Eucalyptus tree of 420 feet, the intermediate form will be, as respects length, some such an herbaceous plant as a St. John's Wort, about 20 inches high. But if we include, as in the case of the animal world, microscopic plants, irrespective of fructification, then, with a small cellular alga, consisting of a single cell, ·01 mm. in diameter, or the one two thousand five hundredth part of an inch as the lowest extreme, we shall have, in longitudinal extension only, the middle place occupied by a small moss, such as *Funaria hygrometrica*, with a total height of less than an inch-and-a-half. In other words, the little moss would be as many times higher than the one cell of the little green alga as the tall tree of the Eucalyptus is higher than the little moss. It would be difficult to calculate bulk for bulk, and estimate size in all directions, so as to ascertain how many such little cells as those of the alga would be required to build up the trunk of such a tree ; but the number would be enormous, so far beyond human experience of numbers that the mind would fail to appreciate their relationship. The intermediate form is larger in plants than in animals, because, although there are animals as small as ·01 mm., there are none reaching 420 feet in height.

Important as are the uses of plants to man, as the source of food, clothing, and medicine, it has hardly been considered as coming within the scope of this volume to refer to them in this aspect, our object being rather to present an accumulation of curious and interesting facts in the structure, habits, or phenomena of plants which have hitherto either been scattered through journals, or presented in the pages of scientific books such as the general public do not take the trouble to read. By this means we have flattered ourselves that we may possibly influence some to take a greater interest in botanical subjects, and in the phenomena of plant life, than they had previously done.

It is a fact worth remembering that vegetation may be conducive to human health and comfort in destroying malaria in pestilential districts. The planting of any "gross feeder" in such places would be beneficial, and the claims which have been advanced on behalf of the Australian gum-tree, might, to a certain extent, be urged on behalf of many other trees. Experience has proved that the planting of any trees which will thrive well and flourish in malarious places, at once produces a marked change for the better, and hygienic plantations need not to be confined to the Eucalyptus. However this may be, it is interesting to note how speedily *Eucalyptus globulus* has found its way into the public newspapers, what patronage it has received in despite of its binomial appellation, and how its cultivation is becoming an article of faith and practice in Europe, Asia, and America, from Rome to Berlin, and from Calcutta to California. There are but few instances on record of a similar vegetable success.

In October, 1873, Mons. Gimbert narrated in "Comptes Rendus"[1] the results of his experiments with the Eucalyptus[2] in Algeria. The tree grows rapidly and aids in destroying the malarious agency which is supposed to cause fever. It absorbs as much as ten times its weight of water from the soil, and emits camphoraceous antiseptic vapour from its leaves. A farm, some twenty miles from Algiers, was noted for its pestilential air in the spring of 1867 : thirteen thousand Eucalyptus trees were planted there, since which time not a single case of fever had occurred. Numerous other like cases are cited.

The honour of discovering this property in the gum-tree is due to Sir W. Macarthur, of Sydney.[3] But this is by no means the only use of the tree. It is valuable as a timber tree. The flowers also yield a large quantity of honey and are much frequented by bees. All parts of the tree are said to be useful as a febrifuge medicine, and the leaves when smoked are efficacious in allaying pain, calming irritation, and procuring sleep. Cigars made of the leaves were

---

[1] "Comptes Rendus," Oct. 6 (1873), p. 764; "Gard. Chron.," Nov. 22, 1873.

[2] *Eucalyptus globulus*.

[3] Naudin on Plantations Hygieniques in "Revue Horticole" (1861), p. 205.

exhibited at the Paris Exhibition of 1867, and recommended as being very efficient in promoting digestion. A chemist at Melbourne also prepares cigarettes from the foliage, which he urges to be employed in bronchial and asthmatic affections. In Mauritius the leaves are sold at sixpence per ounce to make an infusion which has been administered with success in malarious fevers;[1] and, as a reward for all these virtues, as a return for such beneficent work on behalf of humanity, this tree is now being distributed almost over the habitable globe, wherever the white man's foot has trodden.

The sunflower has a reputation similar to that of the Australian gum-tree. "The Observatory at Washington, U. S., was placed in a very unhealthy marshy situation, and at certain periods of the year fever was rife in the neighbourhood, but after the ground was annually sown with sunflower the sanitary condition was much improved." It is also stated by the same authority[2] as that of the above fact, that "a Dutch landed proprietor upon the banks of the Scheldt, planted some plots of sunflowers near his houses, and that the tenants enjoyed afterwards complete immunity from mias-

---

[1] "Lancet," April 20, 1872.
[2] "Gardener's Chronicle," Nov. 22 (1873), pp. 15, 67.

matic fever, although that disease continued to be prevalent in the neighbourhood. In the swampy regions of the Punjab district in India the sunflower is grown in some places in large plantations with marked success, its influence tending to remove malaria, and thus benefit the health of residents in those districts. The Agri-Horticultural Society of the Punjab, after investigating the subject, published a report in which the extension of the cultivation of this useful plant was strongly recommended.

This curious subject would hardly have fallen in its place in any of the subsequent chapters and is therefore alluded to here, in connection with another one to be presently mentioned, rather than be omitted altogether. The influence of vegetation on climate has already received attention in another place,[1] and needs no repetition, although it has an affinity with the facts just referred to. At the same time we might have shown how, and why, such kind of vegetation, as that of the mangrove, aids in perpetrating such a malarious atmosphere as the Eucalyptus is believed to cure. As an illustration of the manner, and the extent, to which the vegetation of a country may be modified and completely changed by external circumstances, we may refer to

---

[1] "Natural History Rambles : The Woodlands," by M. C. Cooke.

South Africa,[1] of which Dr. John Shaw has given a graphic account, the modifying influence being in this case the introduction of the Merino sheep. After alluding to the introduction of a noxious bur-weed (*Xanthium spinosum*), he says that when these sheep were first introduced they fed mainly on grasses, but in a country with periodical rains and a high sun these plants had to give way and succumb. Shrubby plants were not eaten as long as the grass was prominent. But the grass vanished rapidly, and the scrub came to be the main resource of the flocks, and the ground was given over to bush, and scrub, and obnoxious herbs. The climate then became affected, the hardy plants of the southern desert tracts spread northward, and the pleasant country was rapidly becoming an extension of dreary, scrubby, half-deserted Karoo. "Some tracts of the country," he says, "are poisoned by the extraordinary increase of the *Tripteris flexuosa*, and transport riders, with their oxen, our only carrying power, have to travel through certain parts without pausing, on account of the *Melicæ*, grasses which have increased to an extent scarcely to be fancied in the last few years, and on eating which cattle become affected with intoxication to an alarming extent." This is only one example,

---

[1] On the changes going on in the Vegetation of South Africa, in "Linnean Journal," vol. xiv. (1874), p. 202.

out of many which might have been adduced, to show how the surface of the earth is undergoing great modification and alteration, through the disturbing influences of civilisation and colonisation, some of these, such as the destruction of forests, having produced disastrous consequences on the climate.

During 1877, a paragraph went the round of the papers respecting a singular tree, which, although it did not profess to destroy miasma, was no less beneficial, inasmuch as it provided moisture in dry places, and the "Rain-tree," it was anticipated, would convert all deserts into paradise. As there is "nothing new under the sun" the same story, or nearly so, has been found on record more than a century previously, to the following effect: "Near the mountains of Vera Paz (Guatemala) we came out on a large plain, where were numbers of fine deer, and in the middle stood a tree of unusual size, spreading its branches over a vast compass of ground. We had perceived, at some distance off, the ground about it to be wet, at which we began to be somewhat surprised, as well knowing there had been no rain fallen for near six months past. At last, to our great amazement, we saw water dropping, or, as it were, distilling fast from the end of every leaf."[1] The new story, on the authority of

---

[1] "Journey Overland from the Gulf of Honduras," by John Cockburn, London (1735), pp. 40-42.

the United States Consul, related to Moyobamba in Northern Peru, where "the tree is stated to absorb and condense the humidity of the atmosphere with astonishing energy, and it is said that the water may frequently be seen to ooze from the trunk, and fall in rain from its branches, in such quantity that the ground beneath is converted into a perfect swamp. The tree is said to possess this property in the highest degree during the summer season principally, when the rivers are low, and water is scarce, whence it was suggested that the tree should be planted in the arid regions of Peru, for the benefit of the farmers there."

Thus much for the story, as it obtained currency, which requires some modification in face of the facts. The whole subject was investigated, and narrated by Mr. W. T. Thistelton Dyer, in the year 1878.[1] From this we glean the following facts:—The scientific name of the "Rain-tree" was determined as *Pithecolobium saman*. The Director of the Botanic Gardens at Caracas states[2]: "In the month of April the young leaves are still delicate and transparent. During the whole day a fine spray of rain is to be noticed under the tree, even in the driest air, so that the strongly

---

[1] "Nature," February 28 (1878), pp. 349, 350.
[2] Professor Ernst in "Botanische Zeitung" (1876), pp. 35, 36.

tinted iron-clay soil is distinctly moist. The phenomenon diminishes with the development of the leaves, and ceases when they are fully grown." He attributes the rain to secretion from glands on the footstalk of the leaf, on which drops of liquid are found, which are rapidly renewed on being removed with blotting-paper.

Another explanation, furnished by Dr. Spruce, the South American traveller, appears to set the question at rest.[1] "The Tamia-caspi, or Rain-tree of the Eastern Peruvian Andes is not a myth, but a fact, although not exactly in the way popular rumour has lately presented it. I first witnessed the phenomenon in September, 1855, when residing at Tarapolo, a town, or large village, a few days eastward of Moyobamba. A little after seven o'clock we came under a lowish spreading tree, from which, with a perfectly clear sky overhead, a smart rain was falling. A glance upwards showed a multitude of cicadas, sucking the juices of the tender young branches and leaves, and squirting forth slender streams of limpid fluid. My two Peruvians were already familiar with the phenomenon, and they knew very well that almost any tree, when in a state to afford food to the nearly omnivorous cicada, might become a Tamia-caspi, or Rain-tree.

---

[1] "Kew Gardens Report for 1878," pp. 46, 47.

This particular tree was evidently, from its foliage, an Acacia. Among the trees on which I have seen cicada feed, is one closely allied to the Acacias, the beautiful *Pithecolobium saman*. Another leguminous tree visited by cicadas is *Andira inermis*, and there are many more of the same, and other families, which I cannot specify. Although I never heard the name, Tamia-caspi, applied to any particular kind of tree during a residence of two years in the region where it is now said to be a specialty, it is quite possible that, in the space of twenty-one years that have elapsed since I left Eastern Peru, that name may have been given to some tree, with a greater drip than ordinary ; but I expect the cicada will still be found responsible for the 'moisture pouring from the leaves and branches in an abundant shower,' the same as it was in my time."

Although, unfortunately, this explanation takes the romance out of the Rain-tree, it must be admitted that Dr. Ernst is of opinion that the rainy mist in Venezuela is produced without the intervention of insects, and that there is still some mystery to be explained. Under any circumstances, the story is of sufficient interest to warrant an allusion to it in the introduction to the subjects of the present volume, which may contain other phenomena not readily accounted for. Cicadas were great favourites with the ancient Greeks, by whom they were believed to be harmless, and to live upon dew, they were addressed by endearing epithets, and regarded as almost divine.

> Happy creature ! what below
> Can more happy live than thou ?
> Seated on thy leafy throne,
> Summer weaves thy verdant crown ;
> Sipping o'er the pearly lawn
> The fragrant nectar of the dawn.

Plants, regarded in their relationship to different nations and races, have been the theme of more than one writer on botanical geography.[1] There are many suggestions in such a view which are of interest, and we may, in passing, allude to two or three instances. The South Sea Islands are associated with the breadfruit tree, which is the staple food-plant to the natives of Oceania. The lower Coral Islands have the cocoa-nut palm, which grows abundantly in the Indian Islands between Asia and Australia, and on the coasts of India. The New Zealand flax (*Phormium tenax*) is characteristic of the islands from which it derives its name. Amongst the Island Malays we find the clove and nutmeg. Maize was the original possession of the American races. Before the time of the Europeans the maguey plant was the vine of the

---

[1] Schouw, p. 223, etc.

Mexicans, and in recent times another species of the same genus (*Agave Americana*) has acquired the name of Mexican aloe, and furnishes a well-known fibrous material. Above the limit of rye and barley, in Chili and Peru, grows another characteristic plant—the quinoa—the seeds of which are used as food. On the lower Orinoco the savage races subsist on the Mauritia palm. In Africa the date-palm is the inheritance of the Arab. In Abyssinia the coffee appears as the characteristic plant. With the Hindoo it is rice or cotton. In China the tea-shrub is the supreme national plant. Amongst the Indo-Caucasian races of Western Asia and Europe the original characteristic plants are wheat, barley, rye, and oats. Southern Europe has the olive, and, together with Central Europe, the vine. The Laplanders have no characteristic plant, if we except the reindeer moss. Yet all this is being changed with increasing civilisation; as the European races obtained the almond, peach, and apricot from Asia Minor, the orange from China, rice from India, and the maize and potato from America, so the colonies of the same races, established in all climates and scattered over the world, carried with them their characteristic plants, or collected around them those of all other races. In this manner maize, cotton, the vine, coffee, the orange, and even tea, travelling from their original centres, threaten every climate for which they are suitable, and characteristic plants become a legend of the past.

It is scarcely half a century ago since the tea-plant was first introduced for cultivation on the slopes of the Himalayas in India, and now it has become a most important industry; and tea-gardens, formerly unknown, are a distinctive feature in the landscape. More recently, and with similar success, the fever bark, or cinchona plant, has been brought from South America and naturalised on the Neilgherry Hills in Southern India, whence it is spreading to other parts of the Peninsula. To a more limited extent the hop has been introduced from England into the north-west of India, where barley was already grown, and now breweries of "bitter beer" are established for the benefit of Europeans in the most remote regions of our Indian Empire. Not only are useful plants thus widely distributed, but with them others, such as we term "weeds" are associated. The small seeds of these plants, unintentionally mixed with the seeds of food-plants, accompany them to their new destination; thus the red Indian of North America is said to have recognised the plantain, travelling westward with the white man's corn, and gave it the name of the "white man's foot." Every century will make it more difficult of determination what are the really indigenous plants in countries where European races have established themselves.

We may anticipate one or two objections which may possibly be urged against this little volume. One of these may be that we have made very free use of the many researches of Dr. Charles Darwin, in certain phenomena of plant life, without adding to them, in number or in illustration. To this we plead guilty, with the excuse that by so doing we should contribute something towards the diffusion of a knowledge, and, as we hope, of a more general appreciation of the important additions he has made to our knowledge of vegetable life. Some there are who have been content to associate his name only with a theory which they may not comprehend, but do not fail to condemn. With that theory we are not now concerned; but there is another aspect in which we desire that this accurate and indefatigable observer should be known and remembered, outside an exclusively scientific circle; and that is, as a collector of facts, the results of patient observations, illustrative of the life history of plants and animals. The volumes which he has written are unequalled as a cyclopædia of facts; and his bitterest foe has never accused him of distorting, or misrepresenting facts, for the benefit of any theory whatever. As a biological historian, therefore, we commend him to our readers, and, if we have added so little to the subjects which he has investigated, it is because he has done this so completely that further amplification was unnecessary.

The second objection which we may anticipate is the miscellaneous character of the subjects which we have brought together within the two ends of this one book. If the object with which this was undertaken be kept in view, we would fain think that such an objection is also untenable. We profess to be writing a popular volume, on a somewhat unpopular subject. We confess to a design of endeavouring to interest those who are not botanists, and do not pretend to any but a most superficial knowledge of plant life. For such we have collected together, under the headings of a certain number of chapters, a quantity of what we consider curious and interesting phenomena and facts, in the hope that by such means we might stimulate in them an interest in trees, plants, and flowers, which they never felt before. If we succeed in doing this, and, at the same time, in enlarging their views of the power and beneficence of the great Author of all these marvels, our work will have been accomplished.

> Then wherefore, wherefore were they made,
>   All dyed with rainbow light,
> All fashioned with supremest grace,
>   Upspringing day and night;
>
> Springing in valleys green and low,
>   And on the mountains high,
> And in the silent wilderness
>   Where no man passes by?

Our outward life requires them not—
  Then wherefore had they birth?
To minister delight to man,
  To beautify the earth;

To comfort man—to whisper hope,
  Whene'er his faith is dim;
For Who so careth for the flowers
  Will much more care for him.

CHAPTER II.

CARNIVOROUS PLANTS—THE SUNDEWS.

IT is very many years since we wandered about the low swampy parts of Hampstead Heath, in search of the little sundew. It had in those days an interest from its comparative rarity, since it inhabits such localities as are not to be found in every district; but it had also other interest, in the beautiful sparkling glands of the leaves, and its mysterious association with dead insects. This little plant is so inconspicuous that it must be hunted for, amongst the bog moss, in the swampy places in which it delights to grow. The little leaves are nearly as round as a shirt button, and seldom so much as half an inch in diameter, attached at the lower edge to long slender stalks.[1] These stalks radiate from a central point, a short root-stock, and the leaves lie flat on the ground, like a little rosette. In the centre rises the flowering stem, sometimes from four to six inches high, with a few minute white flowers towards the top (fig. 1). The leaves and the ends of the leaf-stalks

---

[1] *Drosera rotundifolia.*

are covered with curious hairs or tentacles, with clubbed ends, which sparkle in the sun, as if they bore on their extremity a minute dew-drop. These leaves, and their curious appendages, are the objects

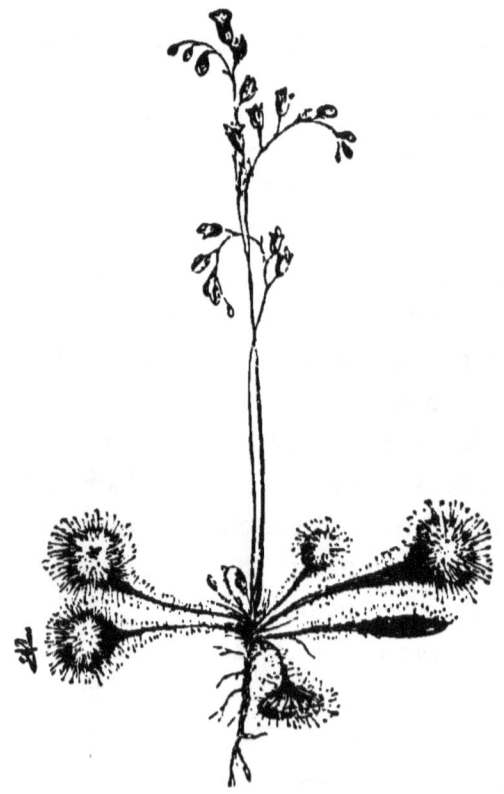

Fig. 1.—Round-leaved Sundew, *Drosera rotundifolia*.

to which our attention must be confined, if we would comprehend why the little sundew has been called a "carnivorous plant."

The leaves, of which the plant seldom bears more than half a dozen, and often less, are covered on the upper surface with glandular hairs, to which the name "tentacles" has been applied. Of these, from 130 to 250 have been counted on single leaves. Those in the centre are shortest and erect, becoming longer and more oblique towards the margin. Each tentacle has a hair-like stem, and bears an expanded oblong gland at the apex. This is surrounded by a viscid secretion, which imparts the glistening dewy appearance that originated the name. If we remove one of these glands, and cut it down the centre, we shall see that it has an external layer of many-sided cells, which are small and filled with purple granular contents (fig. 2). Beneath this is another layer of different-shaped cells, with similar contents. In the centre is a group of clongated cylindrical cells, each with a spiral fibre winding round within it, and containing a limpid fluid. From these spiral cells a spiral vessel runs down through the centre of the stalk or pedicel of the gland. Other and more minute rudimentary hairs are found mixed with the tentacles,

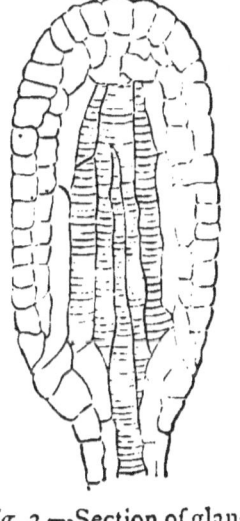

Fig. 2.—Section of gland of *Drosera rotundifolia* magnified.

or covering those parts from which the tentacles are absent.

When any small object is placed on the glands, it causes a movement in the tentacles. The impulse is transmitted from those which are touched to others which surround them, and, one by one, the tentacles bend over towards the centre of the leaf, in order to enclose the irritating object. If the latter is a living object it is more speedily and effectively clasped than a dead one. The time required to cause all the tentacles to close over an object depends upon circumstances. The inflection is more rapid over a thin-skinned insect than a tough-coated one, and the period varies from one to four or five hours for all the tentacles to be closed down upon the captive. If the glands are only touched by a hair or thread, and nothing is left upon them, the tentacles at the margin will curve inwards. This movement may be caused by touching a gland three or four times, and in ten seconds from being touched the movement has been seen to commence.[1]

---

[1] Withering states that in 1780 Mr. Whateley inspected some of these leaves (*D. rotundifolia*) and observed small insects imprisoned therein. On Mr. W. pressing with a pin other leaves, yet in their expanded state, he observed a remarkable sudden and elastic spring of the leaves so as to become inverted upwards, and, as it were, encircling the pin, which evidently showed the method by which the fly came into its embarrassed position.

When an insect is caught by this process, a much more remarkable phenomenon takes place, which was thoroughly examined by Mr. Darwin and declared in the following terms :— " When an object, such as a bit of meat or an insect, is placed on the disc of a leaf, as soon as the surrounding tentacles become considerably inflected, their glands pour forth an increased amount of secretion. I ascertained this by selecting leaves with equal-sized drops on the two sides, and by placing bits of meat on one side of the disc; and as soon as the tentacles on this side became much inflected, but before the glands touched the meat, the drops of secretion became larger. This was repeatedly observed, but a record was kept of only thirteen cases, in nine of which increased secretion was plainly observed ; the four failures being due either to the leaves being rather torpid, or the bits of meat too small to cause much inflection."[1] This is an important fact, as it shows conclusively some relationship between the action of inflection in the hairs and the amount of viscid secretion exuded.

There is, however, another important fact which must be taken into account in connexion with that just recorded. It is, that not only is the secretion increased in quantity, but it also undergoes a change

---

[1] Darwin, " Insectivorous Plants," p. 14.

in its nature, becoming more acid. This acidulation takes place before the glands have touched the object on the leaf, and so long as the tentacles remain bent downwards does the secretion continue to exude, and continues also its acid properties. It might be shown here, as the result of experiment, that fragments of meat, and other substances, placed on the leaves and submitted to the action of this secretion, remained clean and free from putrefaction, whilst other fragments of equal size, placed at the same time on damp moss, became mouldy, or disintegrated, and swarming with infusoria. This fact indicates some preservative power in the acidulated secretion.

It has been demonstrated that most insects are killed within a quarter of an hour from the time of their being caught. The respiration of insects is accomplished by means of breathing pores, or tracheæ, on the surface of their bodies. The viscid secretion from the glands tends to close and choke up these tracheæ, so that the insect is killed by suffocation. Every additional gland, as it closes over the captured insect, contributes of its viscid secretion, which soon bathes and involves the little insect, so that respiration is impossible. The struggles of an insect when first caught only serve to touch and stimulate other tentacles, and increase the number of those which close over it, and pour forth their viscid secretion, and thus hasten its death.

We may well assume, as experiments justify the assumption, that the acidulated secretion, which is discharged over the insect from the inflected glands, aids in the digestion by the plant of this animal food. It is abundantly certain that all these phenomena, the sensibility, or irritability of the tentacles when touched, their power of closing over the object on the leaf, the increase of its viscid secretion, and the acquisition of acid properties, are not performed without a purpose, and *that* purpose appears to be the capture of animal food, its digestion, and ultimate absorption by the plant.

There can be no doubt that the glands of the leaf do really possess the power of absorption, which may be tested by placing upon them small quantities of such substances as carbonate of ammonia, the absorption of which causes a change of colour consequent upon the aggregation of their contents. It may be assumed also from the fact that the tentacles remain closed longer over an object which contains soluble nitrogenous matter than over one which does not. The sundew has very delicate roots, which are scarcely more than suckers for obtaining moisture which the plant requires in great abundance. As Mr. Darwin observes, " a plant of sundew with the edges of its leaves curled inwards, so as to form a temporary stomach, with the glands of the closely inflected tentacles pouring forth their acid secretion, which dissolves animal matter afterwards to be absorbed, may be said to feed like an animal. But, differently from an animal, it drinks by means of its roots; and it must drink largely so as to retain many drops of viscid fluid round the glands, sometimes as many as 260, exposed during the whole day to a glaring sun."[1]

Thus we have taken a cursory glance at the little sundew, and some of the phenomena which it exhibits, in order to comprehend still better the more explicit details of some of the individual features in its history, to which we shall have to return. We have described the leaves, which are in fact the traps by means of which living insects are caught, and, not only this, but the stomach also in which the animal food is digested. To prove that these are not fanciful notions, but have plenty of evidence in support, the important features will have to be examined in detail.

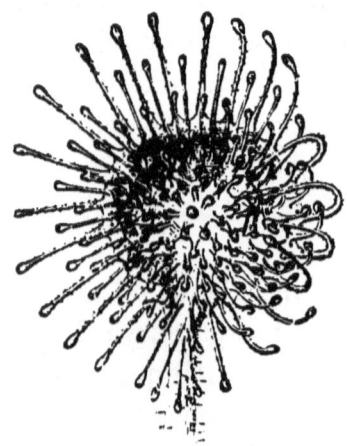

Fig. 3.—Leaf with tentacles of *Drosera rotundifolia*, enlarged.

A leaf studded with sparkling glandular hairs

---

[1] Darwin, "Insectivorous Plants," p. 18.

is not in itself very extraordinary, but when we discover that these hairs, or tentacles, can be moved in a particular direction in response to some exciting cause, we have to deal with a phenomenon by no means common in plant life, and we naturally become curious to discover the cause. When any object, living or dead, comes in contact with one of these tentacles it commences to bend over towards the centre of the leaf (fig. 3). The power of responding to irritation, moreover, is not confined to the single tentacle which has been touched, for it possesses the capacity of communicating with the surrounding tentacles, and they also bend over, as if in sympathy with and to assist their companion. The minute fragment of a human hair $\frac{1}{120}$th of an inch in length, laid upon a gland, has been shown to be sufficient to excite a tentacle to bend over. Minute particles of glass, chalk, and other inorganic substances, placed on the glands of the outer tentacles, will cause them to bend. So also will small fragments of meat, and minute drops of stimulating fluids. When a tentacle is touched three or four times it will also bend, but *not* when only touched once or twice, although the sustained pressure of a gnat's foot is sufficient to produce the movement. After remaining bent down for some time the excited tentacles again slowly return to their original erect position. This return is much more speedy when an inorganic body has been the cause of the inflection, than when a small insect, or a fragment of meat, has been the exciting cause. These facts have been proved by numerous experiments, which place them beyond question. First, that the tentacles are sensitive (if we may use that expression) to the sustained pressure of one millionth part of a grain. That they will respond to such pressure, and bend towards the centre of the leaf. That this irritation will also be communicated to neighbouring tentacles, which will bend in the same direction. And that after this operation is performed the inflected tentacles will return to their former position. If we suppose, then, that a minute insect has fallen or alighted upon one, or more, of the outer tentacles, it will in the course of ten seconds be moving towards the centre, whither it will ultimately be carried, whilst the surrounding tentacles will also follow in the bending movement, until all are closed over the captive insect. But, it may be asked, are we assured that the first inward movement of the tentacles will not alarm the insect and cause it to take flight? It might do so if this were not provided against by the viscid secretion with which the glands are covered, and which increases in quantity with the inward movement of the tentacles. This secretion is so tenacious that it may be drawn out in strings, and if once a small insect alights upon it, it struggles in vain to get free. It is, in fact, a kind of birdlime, prepared naturally, and exposed systematically, for the capture of little flies. The club-shaped summit of each tentacle is a manufactory and storehouse for this sticky substance, which is exuded and exposed on the surface. Although these drops glisten and sparkle in the sun, they have another and more important function to perform than only to justify the cognomen of the plant.

Not only does the response of the tentacles to irritation remind us of sensibility in the animal kingdom, but the apparent power of discrimination which the tentacles possess seems surprising. It is an undoubted fact that the power does exist of distinguishing not only between inorganic and organic substances, as between a piece of glass and a piece of boiled egg, but also between a hard-skinned beetle and a soft fly, and even between different kinds of fluids. Mr. Darwin's experiments give abundant evidence of this, and his book, like all other of his works, is a complete cyclopædia of authentic facts. For instance, drops of pure water were tried on thirty or forty leaves, but no effect whatever was produced. Drops of milk were placed on sixteen leaves, and the tentacles of all became greatly inflected. Ten leaves were tried with drops of cold tea, but the tentacles did not respond. Whereas eight were tested with dissolved isinglass, as thick as milk, and all of them recognised it by inflecting the tentacles. Nor was the treatment of solids less remarkable. "Minute flies were placed on the discs of several leaves, and on others balls of paper, bits of moss and quill, of about the same size as the flies, and the latter (the flies) were well embraced in a few hours, whereas, after twenty-five hours, only a very few tentacles were inflected over the other objects. The bits of paper, moss and quill were then removed from these leaves, and bits of raw meat placed on them, and now all the tentacles were soon energetically inflected."[1] Yet another mode of recognition was manifested. Over and over again, in the work from whence the above is quoted, it is demonstrated that the tentacles remained for a much longer period inflected over what we should term digestible substances than over such indigestible things as bits of glass and paper. The inference to be drawn from this fact is that the plant recognised the latter as indigestible, and hence that the tentacles let go their hold and returned to their previous position of expectancy, whilst in the former they remained closed in the act of digestion. It may be remarked here that as the tentacles, whilst becoming inflected, exude a larger drop of secretion than when erect, so in recovering from inflection they become drier, with little or none of the secretion exuded,

---

[1] Darwin, "Insectivorous Plants," p. 22.

until after they have again resumed their erect position. By the first action the capture and digestion of the prey has to be provided for; by the last any adhering legs or wings of dead insects are got rid of.

We have demonstrated the perfectibility of our fly-catching plant in all that relates to the securing of its prey, and, within certain limits, to its power of selection. The next question is, "What will he do with it?" and this naturally leads us to investigate its powers of digestion and absorption. If the phenomena exhibited by the plant are analogous to those of animals during digestion, we may fairly conclude that the motive is the same. We have stated the fact, which may be repeated in Mr. Darwin's own language, "that the glands of the disc when irritated transmit some influence to the glands of the exterior tentacles, causing them to secrete more copiously, and the secretion to become acid, as if they had been directly excited by an object placed on them. The gastric juice of animals contains, as is well known, an acid and a ferment, both of which are indispensable for digestion, and so it is with the secretion of Drosera. When the stomach of an animal is mechanically irritated it secretes an acid, and when particles of glass or other such objects were placed on the glands of Drosera, the secretion and that of the surrounding and untouched glands was increased in quantity and became acid."[1] It is well known how easy it is to test the presence of an acid by the application of litmus paper, and this test has been applied to the secretion of the glands of the sundew in innumerable instances. The same author says,—" I have tried, indeed, hundreds of times, the state of the secretion on the discs of leaves which were inflected over various objects, and never failed to find it acid." And this observation has been corroborated by others both in this plant and in the Dionœa. When the leaves have not been excited the viscid secretion is not acid, or but very slightly so, but, after the tentacles have commenced bending over any object, the secretion becomes more or less acidulated.

Another property which this secretion possesses has also been alluded to—namely, its antiseptic quality. It checks the appearance of mould and minute animalcules, and for a time prevents the discolouration and decay of substances over which it has been transfused. This, again, is analogous to the gastric juice of animals, which is known to arrest the putrefaction of substances under its influence. And here we have another singular coincidence, even if nothing more, which must have its weight in determining whether the glands of the Sundew possess the power of digestion. If it can be shown that, in addition to

---

[1] Darwin, " Insectivorous Plants," p. 268.

the power of catching insects and holding them, and also of discriminating between digestible and indigestible substances, these leaves secrete a fluid possessing all the attributes of a digestive fluid, dissolving without putrefaction just such substances as an animal would dissolve in its stomach by the ordinary process of digestion, we furnish very strong presumption in favour of their being called " insectivorous."

Although his remarks were illustrative of another plant, we may better quote here the observations of Dr. Burdon Sanderson, as they apply with equal force to the sundew as to the Venus's fly-trap. In his lecture at the Royal Institution,[1] after describing the plant and its mechanism, he referred to its power of digestion. "When," he says, "we call this process digestion we have a definite meaning. We mean that it is of the same nature as that by which we ourselves, and the higher animals in general, convert the food they have swallowed into a form and condition suitable to be absorbed, and thus available for the maintenance of bodily life. We will compare the digestion of Dionœa with that which in man and animals we call digestion proper, the process by which the nitrogenous constituents of food are rendered fit for absorption. This takes place in the

---

[1] June 5th, 1874; reported in "Gardener's Chronicle" for June 27th, 1874.

stomach. It also is a fermentation, *i.e.*, a chemical change, effected by the agency of a leaven or ferment which is contained in the stomach juice, and can be, like the ferment of saliva, easily separated and prepared. As so separated it is called pepsin. Consequently, having the ferment, we can easily imitate digestion out of the body. For this experiment there are three things necessary (1) That our liquid should contain pepsin ; (2) That it should be slightly acid ; (3) That it should be kept at the temperature of incubation (about 97° Fahr.). We select for the experiment a substance which, although nutritious and containing nitrogen, is not easily digested—such, for example, as boiled white of egg. In water containing a small percentage of hydrochloric acid, and a trace of pepsin, it is gradually dissolved ; but chemical examination of the liquid shows us that it has not been destroyed, but merely transformed into a new substance called peptone, which is afterwards absorbed, *i.e.*, taken into the circulating blood."

"Between this process and the digestion of the Dionœa leaf the resemblance is complete. It digests exactly the same substances in exactly the same way, *i.e.*, it digests the albuminous constituents of the bodies of animals just as we digest them. In both instances it is essential that the body to be digested should be steeped in a liquid, which in Dionœa is secreted by the red glands on the upper surface of the leaf; in the other case by the glands of the mucous membrane. In both the act of secretion is excited by the presence of the substance to be digested. In the leaf, just as in the stomach, the secretion is not poured out unless there is something nutritious in it for it to act upon; and, finally, in both cases the secretion is acid. As regards the stomach we know what the acid is,—it is hydrochloric acid. As regards the leaf we do not know precisely as yet, but Mr. Darwin has been able to arrive at very probable conclusions."

It has been demonstrated, by experiment, that the secretion of the glands of the sundew completely dissolves albumen, muscle, fibrin, cartilage, the fibrous portion of bone, gelatin, and the casein of milk. That is to say, little cubes of hard-boiled egg, fragments of roast meat, tough cartilage from a leg-bone of mutton, small pieces of the bone of a fowl, and of a mutton-chop bone, the latter being so softened that it might be penetrated by a blunt needle, or compressed, and these became dissolved as they would have been in the stomach of some of the higher animals.

It is only necessary to cite one experiment which was performed on a most unpromising substance under somewhat unfavourable conditions. "Three cubes of white translucent, extremely tough cartilage were cut from the end of a slightly-roasted leg-bone of a sheep. These were placed on three leaves, borne by poor small plants in my greenhouse during November; and it seemed in the highest degree improbable that so hard a substance would be digested under such unfavourable circumstances. Nevertheless, after forty-eight hours, the cubes were largely dissolved, and converted into minute spheres, surrounded by transparent, very acid fluid. Two of these spheres were completely softened to their centres, whilst the third still contained a very small irregularly-shaped core of solid cartilage. Their surfaces were seen under the microscope to be curiously marked by prominent ridges, showing that the cartilage had been unequally corroded by the secretion. I need hardly say that cubes of the same cartilage, kept in water for the same length of time, were not in the least affected."[1]

The fact, therefore, is clearly established, that the secretion from the glands of the Sundew, under certain conditions of stimulation, is capable of dissolving animal substances, in precisely the same manner as they are acted upon during the process of digestion in the stomach of animals. It remains to be seen what evidence there is in support of the absorption, and assimilation, of the substances so digested. Here it would be essential to show, in the first instance, that the glands in question possess the power of absorp-

---

[1] Darwin, "Insectivorous Plants," p. 104.

tion at all. Because if it can be demonstrated that they are capable of absorbing fluids, especially nitrogenous fluids, it would be easy to believe that no exception would be made to the exclusion of dissolved animal substances.

All the experiments made in this direction are exceedingly interesting and instructive, some of them truly marvellous in their results. It would be somewhat tedious to narrate them in detail, in a popular exposition of the reasons why certain plants have been called "carnivorous plants," but it will be necessary to allude to one or two. Solutions of certain chemical substances, called salts of ammonia, were applied to the leaves of living plants. Some of these quickly discoloured the glands, but all caused the characteristic inflection of the tentacles. Yet these salts were applied in a very diluted state, for less than one-millionth part of a grain, absorbed by a gland of one of the exterior tentacles, was sufficient to cause it to bend. In order that some idea might be formed of what a million means, the following illustration is given in a foot-note to Mr. Darwin's book. "Take a narrow strip of paper, eighty-three feet four inches in length, and stretch it along the wall of a large hall; then mark off at one end the tenth of an inch. This tenth will represent a hundred, and the entire strip a million." The experiments alluded to were performed in three ways.

Small drops were placed on the disc of the leaves; very minute drops were gently placed on one or more of the exterior tentacles; and whole leaves were cut off and immersed in the solutions. In all ways the results harmonised. How exceedingly sensitive the leaves were to some of these solutions may be inferred from their great dilution. As an illustration, it was stated that five thousand fluid ounces would more than fill a thirty-one gallon cask, and that to this large body of water one grain of the salt was added; only half a drachm, or thirty minims, of the solution being poured over a leaf. "Yet this amount sufficed to cause the inflection of almost every tentacle, and often of the blade of the leaf."

The solution of many salts, acids, and alkaloids, were tried, and produced in some instances unexpected results, inasmuch as some substances which are wholly harmless to animals were poisonous to the plant, and others which are poisonous in their effects upon animals were almost inert upon the Sundew. Abundant evidence was supplied that the fluids must have been absorbed by the glands, and their influence transmitted to other tentacles which were not touched. In cases of poisoning, for instance, it must be conceded that the deleterious substance was absorbed, for the parts became blackened, and all the phenomena of poisoning were exhibited. It is but fair to conclude that, if deleterious substances actually become absorbed, the result of absorption being plainly traced, that also other substances may be absorbed, which would either be neutral or beneficial to the plant, but which cannot so easily be traced in their course.

There is another remarkable phenomenon in which the tentacles perform a conspicuous part, which must be briefly alluded to, as affording evidence of the great changes which take place in the internal organism of these plants under excitement. If a resting tentacle is examined, selecting one which has not been excited or inflected, the cells of the pedicel will be seen to present a completely uniform appearance, filled with a purple fluid, retaining throughout one uniform character, and a thin layer of uncoloured circulating fluid, passing along the walls of the cell. These cells, with a diffused colour, impart to the pedicel a continuous purple tint. But if a tentacle is examined after it has been excited, from whatever cause, the cells will present quite a changed appearance. Even to the naked eye they will not present the same uniform, even, purple tint, but seem to be speckled, mottled, or variegated. Examination of these cells, so changed, under the microscope will reveal the cause of the mottling, in the aggregation of the purple matter, which they contain, in variously-shaped masses, suspended in a colourless medium. Each cell, which before was suffused with a uniform tint, now holds a clear and colourless fluid, in which floats an elongated dark-coloured body, formed by the aggregation of the colouring matter. This aggregated body, sometimes of a single mass, sometimes of two, is constantly changing its form, slowly but gradually, like that curious little animal found in stagnant waters called an Amœba. Finally, the movement ceases, the masses again dissolve, and become diffused through the contents of the cell, which again assume a uniform tint and appearance. In this cycle we have a manifestation of a great molecular change which is wrought within the cells of the tentacles, in response to some external irritation. Whatever causes the tentacle to become inflected seems also sufficient to induce this phenomenon of the aggregation of masses within the cells. When the influence of that irritation has passed away, and the tentacle has assumed its original erect position, the contents of the cells assume also their homogeneity, or uniform density.

This aggregation commences in a tentacle at the upper end, in immediate proximity to the gland, and proceeds from above downwards. It accompanies the bending over of the tentacle, but that it neither causes the inflection, nor is caused by it, is evident from the fact that aggregation may take place when there is no inflection of the tentacle. Some acids will produce a rapid inflection but no aggregation.

Hence, then, this aggregation is neither a cause nor a consequence of inflection. Whatever its cause may be, it appears to be invariably accompanied by an increased secretion of the glands, and the dispersion of the masses, in like manner, indicates a diminution in the amount of viscid matter secreted by the glands. The shorter central tentacles, which have a green pedicel, exhibit the same phenomenon, with the exception that the aggregated masses partake of the green colour of the cells. The colour of the aggregated masses being of course dependent upon the colour of the contents of the cells.

The experiments on the Sundews have been for the most part conducted with the little round-leaved Sundew, but the other two English species have likewise been examined, and found to correspond in all essentials with its fellow species. This has been done both in England and America. A species with very long slender leaves,[1] which grows abundantly in New Jersey, has been tested in a similar manner. One person writes:[2] "I found it in full bloom, and growing as thick as it could well stand, on either side of an extensive cranberry plantation. This charming plant with its pretty pink blossoms, together with the dew-like substance exuding from the glands (the glands

---

[1] *Drosera filiformis.*
[2] Mrs. Mary Trent in "American Naturalist," vii., Dec., 1873.

surmount the bristles or hairs which cover the long thread-like leaves) was one of the most beautiful sights I ever beheld. From former observations I had supposed this plant caught only small insects, but now found I was mistaken; great Asilus flies were held firm prisoners, innumerable moths and butterflies, many of them two inches across, were alike held captive until they died—the bright flowers and brilliant glistening dew luring them on to sure death. But what is the use of this wholesale destruction of insect life? can the plants use them? Upon examination I find that after the death of the larger insects, they fall around the roots of the plants, and so fertilise them, but the smaller flies remain sticking to the leaves."

And again, "At ten o'clock I pinned some living flies half an inch from the leaves, near the apex. In forty minutes the leaves had bent perceptibly toward the flies. At twelve o'clock the leaves had reached the flies, and their legs were entangled among the bristles and held fast. I then removed the flies three quarters of an inch further from the leaves. The leaves still remained bent away from the direction of the light toward the flies, but did not reach them at this distance."

Mr. Darwin also examined this species, which he says, "had thread-like leaves from six to twelve inches in length, with the upper surface convex and the lower flat and slightly channelled. The whole convex surface down to the roots—for there is no distinct footstalk—is covered with short gland-bearing tentacles, those on the margin being the longest and reflexed. Bits of meat placed on the glands of some tentacles caused them to be slightly inflected in twenty minutes, but the plant was not in a vigorous state." Two Australian species have also exhibited the same propensities, so that it is probable that all the Sundews, from whatever part of the world they may come, are equally fly-catchers, as well as our own species continue to be, when found flourishing in countries far remote.[1]

---

[1] As, for instance, in Trinidad, where it was observed by the Rev. Charles Kingsley:—"As I scratched and stumbled along the tussocks, 'larding the lean earth as I stalked along,' my kind guide put into my hand, with something of an air of triumph, a little plant, which was—there was no denying it— none other than the long-leaved Sundew, with its clammy-haired paws full of dead flies, just as they would have been in any bog in Devonshire or in Hampshire, in Wales or in Scotland. But how came it here (in Trinidad)? And, more, how has it spread, not only over the whole of Northern Europe, Canada, and the United States, but even as far south as Brazil? Its being common to North America and Europe is not surprising. It may belong to that comparatively ancient flora which existed when there was a landway between the two continents by way of Greenland, and the bison ranged from Russia to the Rocky Mountains. But its presence within the tropics is more probably explained by supposing that it has been carried on the feet or in the crop of birds."—" At Last," p. 315.

In Dr. Darwin's book all the facts resulting from observation, and experiment are brought to bear upon the theory which he advanced, that the power of catching and digesting insects is of advantage to the plants themselves. Some continental botanists have denied that the case is proved. Subsequent to the volume in question Mr. Francis Darwin instituted some experiments with the view of ascertaining what effect the indulgence in carnivorous propensities had upon the Sundew. The plants were isolated and protected. Half the plants, or 91 plants, were not fed, whilst 86 plants were supplied with roast meat, cut into thin slices across the grain, and the fibre torn into fragments exceedingly minute. The first difference observed was that in August the starved plants had only produced 116 flowering stems whilst the fed plants had produced 173. Another difference observed was that the fed plants contained a larger number of healthy leaves than the starved plants, and, finally, it is stated that from these experiments "it would seem that the great advantage accruing to carnivorous plants from a supply of nitrogenous food to the leaves is the power of producing a vastly superior yield of seeds; and," the author adds, "I venture to think that the above experiments prove beyond question that the supply of meat to Drosera is of signal advantage to the plants." Similar experiments in Germany in which the plants were fed with plant-lice instead of meat resulted in a similar conclusion, that numerous and striking advantages accrued to the fed plants.[1]

Admitting that the case is not sufficiently proved for us confidently to affirm that these carnivorous habits are conducive to the welfare of the plant, we cannot deny its probability, because otherwise we are placed in the dilemma of assuming, either that all this adaptation for catching and destroying animal life is wanton mischief, or that it is an expenditure of power without purpose. From experience of the operations of nature we are unwilling to recognise such a departure from the usual plan. We are accustomed to trace operations performed by an economy of force, and to believe that nothing is done in vain. Wanton destruction, or wasted energy, are not the probabilities which would suggest themselves to the mind of any one who has devoted himself to the study of the phenomena of life, nor would they elevate our conception of the All-wise Creator, of what in such a case would be undoubted failures.

---

[1] "Journal of Linnean Society" (Botany), xvii., pp. 17 to 32.

CHAPTER III.

CARNIVOROUS PLANTS—VENUS'S FLY-TRAP.

BELONGING to the same natural order of plants as the Sundews, Venus's Fly-trap, or, botanically, *Dionæa muscipula*, has recently been much harassed by experiments to test its flesh-eating capacity. It is not a British native, but an inhabitant of damp places in the eastern parts of North Carolina, so that its relationship to our Sundew may be described as that of an "American cousin." In like manner it will grow and flourish in wet moss, without any soil, and consists of a rosette of leaves, which radiate from a centre, but both leaves and tufts are larger, and more conspicuous, than in the Sundew. The foot-stalk of the leaves is flattened out, and leaf-like. The blade of the leaf is somewhat rounded in outline, and composed of two lobes, which are hinged down the centre, so that the lobes rise up, and apply themselves together face to face. Around the margin of the lobes stands a row of bristles, which will be more fully described shortly. With a coloured figure of this plant, published ninety years ago in "Shaw's

Miscellany," is the following remark: "The surface of the leaves is irritable in the highest degree, and whatever insect is so unfortunate as to alight on it is caught as effectually as a mouse in a trap, and is even generally squeezed to death by the pressure. What particular purpose in the economy of nature is answered by the imprisoning power of this extra-

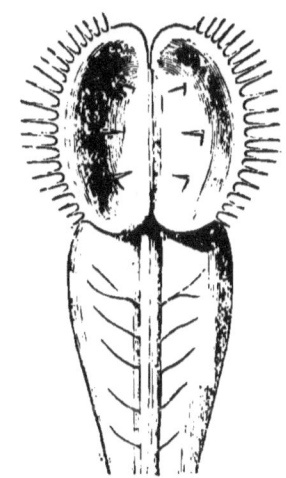

Fig. 4.—Venus's Fly-trap, *Dionœa muscipula*.

ordinary vegetable, it is extremely difficult, and perhaps impossible, to determine."

As long ago as 1768, a naturalist, named Ellis, called the attention of Linnæus to the peculiarities of the leaves of the Venus's Fly-trap, or Dionœa, by the following remarks: "The plant shows that Nature may have some views towards its nourishment, in forming the upper joint of its leaf like a machine to catch food; upon the middle of this lies the bait for the unhappy insect that becomes its prey. Many minute red glands that cover its surface, and which perhaps discharge sweet liquor, tempt the poor animal to taste them; and the instant these tender parts are irritated by its feet, the two lobes rise up, grasp it fast, lock the rows of spines together, and squeeze it to death. And, further, lest the strong efforts for life in the creature just taken should serve to disengage it, three small erect spines are fixed near the middle of each lobe, among the glands, that effectually put an end to all its struggles. Nor do the lobes ever open again while the dead animal continues there. But it is, nevertheless, certain that the plant cannot distinguish an animal from a vegetable or mineral substance; for if we introduce a straw or pin between the lobes it will grasp it full as fast as if it were an insect." Linnæus, however, only regarded these phenomena as illustrations of the extreme sensibility of the leaves. Sixty years subsequently, Dr. Curtis, of North Carolina, made further and more complete examination of these leaves. "Each half of the leaf," he says, "is a little concave on the inner side, where are placed three delicate hair-like organs, in such an order that an insect can hardly traverse it without interfering with one of them, when the two sides suddenly collapse and enclose the prey with a force surpassing an insect's efforts to escape. The fringe of hairs of the opposite sides of the leaves interlace, like the fingers of two hands clasped together. The sensitiveness resides only in these hair-like processes on the inside, as the leaf may be touched or pressed in any other part without sensible effects." After this, another American botanist, who was staying in the district where the " fly-trap " flourishes, resolved upon some experiments, and by feeding the leaves with small pieces of beef he found that these were completely dissolved and absorbed; the leaf opening again with a dry surface, and ready for another meal, though with an appetite somewhat jaded. He found that cheese disagreed horribly with the leaves, turning them black and finally killing them.[1]

The insectivorous predilections of *Dionœa* have, therefore, been suspected, if not demonstrated, for more than a century. In the account of the plant given by Shaw, he commences by alluding to the different methods by which carnivorous animals catch their prey, and then he adds, " What is still more extraordinary, there are not wanting amongst vegetables some instances in which the smaller animals meet their fate by alighting on the flowers or leaves; being either held fast by a viscous exudation from

---

[1] " Gardener's Chronicle," August 29th, 1874, p. 260.

the surface or confined by the pressure of the irritable parts of the plant." The date attached to the plate is 1790.

Sir Joseph Hooker[1] has given a complete summary of the history of all the observations which have been made on the plant from the earliest times, which may be consulted by any who desire a more explicit narrative of the details than our space will enable us to furnish. We direct ourselves at once to the *modus operandi* by means of which the plant achieves its object. This mechanism has been compared by Dr. Burdon Sanderson to a rat-trap. "When it (the leaf) is open, the lobes are at right angles to each other. When an insect comes into contact with either, at once they approach each other, but this does not occur with the suddenness and completeness that it occurs in a rat-trap. The lobes begin to close sharply enough, but do not come quite together, remaining for some time *entr'ouvert*. When the leaf is in this state of half-closure, it is easy to see what is the significance of the two sets of prongs. You see that they are set on alternately, along the opposite edges of the lobes, so that, just like the teeth of a rat-trap, they fit into each other. It is not difficult to see why this is so, *i.e.*, why the spikes are arranged alternately. The

---

[1] Address to the British Association at Belfast in 1874.

leaf, being a trap, is made like a trap. But I should not have been able to tell you why the leaf does not at once close on its prey had not Mr. Darwin told me. After having partially closed, as I have said, one of two things may happen. The insect, having been caught, at once begins to think of escaping, and makes efforts to do so, which may or may not be successful. If it is small, it easily finds its way out through this wonderful grating formed by the crossing of the teeth, and in this case the leaf soon recovers, expands again, and is ready for the capture of another victim. If it is large, all its efforts to regain its liberty are futile. Repelled by its prison-bars, it is driven back upon the sensitive hairs which stick into the interior of its cell, and again irritates them. By doing so it occasions a second and more vigorous contraction of the lobes. The result is, that the creature is not only captured, but crushed ; not only swallowed, but digested."

The minute structure of the leaves differs in many respects from that of the Sundews. The rigid marginal spines are without glands on their tips, and are not irritable. The three minute filaments which project from the upper surface of both lobes, on the contrary, are remarkable for their extreme sensitiveness to the touch, but they also are pointed at their extremity. Besides these cuticular appendages the upper surface is thickly covered, except near the margin, with minute reddish or purplish glands, but there are no glands on the leaf-like foot-stalk. The glands are elevated on short pedicels, and are convex above. Little stellate projections of an orange-brown colour, with eight radiating arms, are scattered over the foot-stalk, the back of the leaves, and the basal part of the marginal spikes, and a very few on the surface of the lobes. Here and there a few minute pointed hairs may be traced on the back of the leaves.

The functions of all these parts have been fairly ascertained. That of the marginal spines is of a mechanical nature, and perhaps entirely so, as they are neither sensitive nor glandular, and do not seem to possess any separate or spontaneous motion. The sensitive filaments, on the contrary, are eminently sensitive. Their apices are sometimes divided into two or three points, and, from apex to base, it is impossible to touch them, ever so lightly, without at once acting on the lobes of the leaf and causing them to close. These sentinel filaments, although so sensitive to a slight touch, are less sensitive to prolonged pressure. This difference between the filaments in Dionœa and the glands of Drosera relates to the different habits of the two plants. It has been seen how a slight prolonged pressure acts on the Sundew; but in the Dionœa there is no viscid secretion to detain the insect, which must be caught at once by the rapid closing of the lobes, simultaneously with the slightest touch; for the filaments neither secrete nor absorb, and are, in fact, purely sentinels. The tentacles of Drosera when excited become inflected and aggregated, but this property does not extend to the Dionœa filaments. Drops of water falling on them will not cause the lobes to close, nor blowing upon them strongly. Hence the sentinels are not likely to give a false alarm at a shower of rain or a gale of wind. Neither did the rays of the sun, when concentrated upon the filaments to such a degree as to cause them to be scorched and discoloured, produce any movement.

The minute glands with which the surface of the leaves is studded have the power of secretion and absorption, but they do not secrete until excited by the presence of animal matter. Other objects placed upon the glands will remain quite dry; but, if a fragment of meat, or a crushed fly, is placed on the surface of the expanded lobes after a time the glands will secrete freely. If the lobes are made to close over an insect, then the glands of the whole surface secrete copiously. Two or three instances are given by Mr. Darwin in proof of this:[1] "On one occasion when a leaf was cut open, on which a small cube of albumen had been placed forty-five hours before,

---

[1] Darwin, " Insectivorous Plants," p. 296.

drops rolled off the leaf. On another occasion in which a leaf, with an enclosed bit of roast meat, spontaneously opened after eight days, there was so much secretion in the furrow over the midrib that it trickled down. A large crushed fly was placed on a leaf from which a small portion at the base of one lobe had previously been cut away so that an opening was left, and through this the secretion continued to run down the foot-stalk during nine days—that is, for as long a time as it was observed."

Aggregation, which was insisted upon in our remarks on the Sundews, may be seen to take place very quickly in the glands of the Dionœa, after contact with nitrogenous subjects, every cell having its contents aggregated, in a beautiful manner, into dark, or pale purple, or colourless, globose masses of protoplasm. The function of the little stellate projections, with eight radiating arms, not having been demonstrated can only be conjectured.

From these details of the structure of the leaves we are enabled to correlate them with their movements. When an insect touches one of the sentinel filaments, on an expanded leaf, the irritation is at once communicated, and the lobes close together, with the captured insect enclosed between them, its struggles, in so far as they touch the filaments, only serving to accelerate the closing. The interlocking marginal spines prevent any escape, except in the case of very minute insects. Contact with the glands causes them to absorb, and then the secretion of the acid-mucilaginous fluid commences, and proceeds as long as any material is left to stimulate the action of the glands. Under this treatment the insect becomes dissolved, as far as it is capable of dissolution, and is assimilated by the leaf, this action causing aggregation of the protoplasm in the cells of the glands. All these steps in the process have been determined, by means of careful experiment, which we have not deemed it necessary to recount. That the captured insects were in some way made subservient to the nourishment of the plant was conjectured from the first. Dr. Curtis found them enveloped in a fluid of mucilaginous consistence,[1] which seemed to act as a solvent, the insects being more or less consumed by it. This was verified, and the digestive character of the liquid well-nigh demonstrated some years ago by Mr. Canby, of Wilmington, who, upon a visit to North Carolina, and afterwards at his own home, followed up Dr. Curtis's suggestions with some capital observations and experiments, which were published in 1868,[2] although they did not seem to have attracted the attention which they deserved.

---

[1] Dr. Curtis in "Journal of Boston Society of Natural History," vol. i., 1834.

[2] Canby in "Meehan's Gardener's Monthly," vol. x., August, 1868 (Philadelphia).

The points which Mr. Canby made out are, that this fluid is always poured out around the captured insect in due time; "if the leaf is in good condition and the prey suitable"; that it comes from the leaf itself, and not from the decomposing insect (for, when the trap caught a plum curculio, the fluid was poured out while he was still alive, though very weak, and endeavouring, ineffectually, to eat his way out); that bits of raw beef, although sometimes rejected, after awhile were generally acted upon in the same manner—*i.e.*, closed down upon tightly, slavered with liquid, dissolved mainly, and absorbed; so that, in fine, the fluid may well be said to be analogous to the gastric juice of animals, dissolving the prey, and rendering it fit for absorption by the leaf. Many leaves remain inactive, or slowly die away, after one meal; others re-open for a second, and perhaps a third capture, and are at least capable of digesting a second meal.

When the lobes close together from irritation by inanimate substances, or touching, the inner surface remains concave until the lobes expand again; but, if an insect or a piece of meat is enclosed, each lobe gradually flattens, and that apparently with considerable force, thus pressing the enclosed object firmly against the secreting glands. When no object is caught the lobes soon expand again in from twenty-four to thirty-two hours, and even before fully expanded they are ready to act again, so that a leaf has been found to close and re-open alternately, but unsuccessfully, for four times during six days. Closing in this manner, from irritation by inanimate objects, does not, therefore, prevent the lobes from acting vigorously several times, until some suitable prey is caught, and then they remain closed for an indefinite period, or, if they open again at all, remain torpid and insensible for a considerable period. " In four instances leaves after catching insects never re-opened, but began to wither, remaining closed—in one case for fifteen days over a fly; in a second, for twenty-four days, though the fly was small; in a third, for twenty-four days over a woodlouse; and in a fourth, for thirty-five days over a crane fly. In two instances, in which very small insects had been naturally caught, the leaf opened as quickly as if nothing had been caught."[1] Dr. Canby says that the leaves remain closed for a longer period over insects than over meat.

In all cases where the leaves re-opened, after having remained a long time closed over insects, or meat, or similar substances, they were so torpid during many succeeding days that touching the sensitive filaments was followed by no response whatever. In their native country, where the plants grow with vigour,

---

Darwin, " Insectivorous Plants," p. 309.

they appear to be more capable of repeating their operations than when transplanted here. Mrs. Trent, who cultivated and watched these plants in New Jersey, which is not so far removed from their natural habitat, has stated that "several leaves caught successively three insects each, but most of them were not able to digest the third fly, but died in the attempt. Five leaves, however, digested each three flies, and closed over the fourth, but died soon after the fourth capture. Many leaves did not digest even one large insect." The capacity for digestion is not, therefore, unlimited in the Dionœa, more than it is in higher organisms. Apoplexy from over-feeding might even here be a reasonable verdict.

As to the kind of insects which are captured by this plant we have the record of the contents of fourteen leaves, sent, with their prey, from their native country.[1] Four of these had caught rather small insects, of which three were ants, and the fourth a small fly, but the other ten had caught large insects, of which eight were beetles (two chrysomelas, five claters, and a curculio), a thick broad spider, and a scolopendra. Of the whole there was only one flying insect, or, rather, usually and readily progressing by flight. This hardly seems to harmonise with the statement by Dr. Canby that "as a general thing beetles and

---

[1] Darwin, p. 312.

insects of that kind, though always killed, seem to be too hard shelled to serve as food, and after a short time are rejected."[1]

We may here allude to that phase of the subject which was so successfully investigated and illustrated by Dr. Burdon Sanderson, and which amounted to establishing the identity of the phenomena of muscular contraction and contractility in Dionœa. The property of contracting when irritated, which enables the Dionœa to catch insects, was the special phase of the subject to which Dr. Burdon Sanderson directed his attention. In this phenomenon, he says, " we have to do not merely with contractility but with irrato-contractility. The fact that the property requires two words to express it implies that there are two things to express, viz. (1) that contraction takes place, and (2) that it takes place in answer to irritation. As this is the case, not only here, but in all other instances of animal or vegetable active motion, we recognise in physiology these two properties as fundamental—irritability or excitability, and contractility, the former designating the property, possessed by every living structure whatever, of being excited into action (*i.e.*, of having its stored-up force discharged) by some motion or disturbance from outside; the latter, that kind of discharge or action which

---

[1] " Gardener's Monthly," August, 1868.

results in change of form, and usually declares itself in the doing of mechanical work. This property of excitability, which, let me repeat, is common to all living structures, is, as we have seen, comparable in its simplest manifestations to that possessed by many chemical compounds (of explosiveness) and many mechanical contrivances (of going off or discharging when meddled with, as in the case of the rat-trap already referred to).

"In physiology, as in the other sciences of observation, the process of investigation is throughout one of comparison. Not only do we proceed, from first to last, from the known towards the unknown, but what we speak of as our knowledge, or understanding, of any new fact consists simply in our being able to bring it into relation with other facts previously well ascertained and familiar, just as the geographer determines the position of a new locality by ascertaining its topographical relation to others already on the chart.

"The comparison we have now to make is between the contractility displayed by the leaf of Dionœa, and the contractility of muscle. I choose muscle as the standard of comparison, because it is best known, and has been investigated by the best physicists of our time, and because its properties are easily illustrated and understood. I shall be able to show that the resemblance between the contraction of muscle and that of the leaf is so wonderfully complete, that the further we pursue the inquiry the more striking does it appear. Whether we bring the microscope to bear on the structural changes which accompany contraction, or employ the still more delicate instruments of research, which you have before you this evening, in order to determine, and measure, the electrical changes which take place in connexion with it, we find that the two processes correspond in every essential particular so closely, that we can have no doubt of their identity.

"Muscle, like every other living tissue, is the seat, so long as it lives, of chemical changes, which if the tissue is mature, consist entirely in the disintegration of chemical compounds and the dissipation of the force stored up in these compounds, in the form of heat or some other kind of motion. This happens when the muscle is at rest, but much more actively when it is contracting, in which condition it not only produces more heat than it produces at other times, but also may do—and under ordinary circumstances does—mechanical work; these effects of contraction of muscle are, of course, dependent in quantity on the chemical disintegration which goes on in its interior.

"Again, muscle, so long as it is in the living state, is electromotive. This property it probably possesses in common with other living tissues, for it is very likely that every vital act is connected with electrical change in the living part. But in muscle, as well as other irritable and contractile tissues in animals, the manifestation of electromotive force is inseparably connected with the special function of the tissue *i.e.*, with contraction, the connexion being of such a nature, that the electromotive force expresses, not the work actually done at any given moment, but the capacity for work. Thus, so long as the muscle lives, its electromotive force is found to be on the whole proportioned to its vigour. As it gradually loses its vitality, its power of contracting and its electromotive force disappear *pari passu*. When it contracts, the manifestation of electromotive force diminishes in proportion to the degree of contraction. But it is to be borne in mind that, although, when the muscle or the leaf contracts, electromotive force disappears and work is done, there is no reason for supposing that there is any conversion of the one effect into the other, or that the source of the force exercised by the organ in contracting is electrical."

Dr. Burdon Sanderson then proceeded by a series of experiments to demonstrate the correspondence between the electrical phenomena which accompany muscular contraction and those which are associated with the closing of the Dionœa leaf.[1]

---

[1] Lecture by Dr. Burdon Sanderson at Royal Institution, June 5th, 1874; "Gardener's Chronicle," June 27th, 1874.

With this brief and rapid summary of the main features relating to the carnivorous propensities of the Venus's Fly-trap, we may casually refer to a few other plants belonging to the same natural order as the Sundews and Dionœa, which possess similar propensities, but to a less interesting degree, or do not differ greatly from the two preceding types.

A little aquatic plant, called *Aldrovanda vesiculosa* is found in Europe, Australia, and India. Although inhabiting countries so remote from each other, this plant seems to be of one species in all. It has no roots, and floats like green stars in the water. The leaves are arranged in whorls in a stellate manner round the stem. Each leaf has two semicircular lobes, which are seated on broad foot-stalks. The lobes are generally found closed at the ordinary temperature in Europe, but they *do* separate, under favourable conditions, to about the same proportionate extent as a living mussel opens the valves of its shell.

The history and mystery of this little water-plant are very imperfectly known. Stein observed that water insects were sometimes caught by it. Professor Cohn has found crustaceans and larvæ within the leaves.[1] Plants placed in water containing entomostraca were examined next morning, and found to enclose individuals of these minute crustaceans still alive. In

---

[1] Cohn, "Beitrage," iii., 1875, p. 71.

one of the closed leaves of the Australian variety from Queensland a rather large beetle was found, with all the softer parts of the body dissolved.

The leaves evidently are well adapted for catching living creatures. There are long sensitive hairs which are probably sensitive. There are glands which, from analogy, may secrete a limpid fluid.[1] Altogether, however, although kinship and analogy might point to this as another of the carnivorous plants of the Sundew family, a supposition which is supported by a sort of circumstantial evidence, still, so little is definitely known, that it is better to suspend the judgment than reach at too hasty a conclusion.

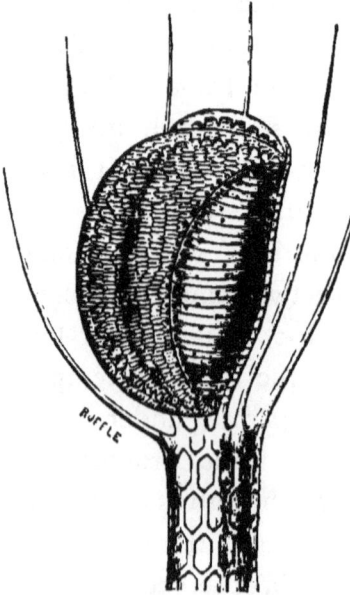

Fig. 5.—Leaf of *Aldrovanda*, enlarged.—COHN.

The Portuguese Fly-catcher is the name by which we may distinguish that rare little plant *Drosophyllum Lusitanicum*, which hitherto has only been found in Portugal and Morocco. It is plentiful in the neigh-

---

[1] Darwin, p. 330.

bourhood of Oporto, where the villagers call it the "fly-catcher," and hang it in their cottages for that purpose. The leaves are like slender filaments, of several inches in length, with the upper surface concave and channelled down the middle, and the under surface convex. Both surfaces are covered with tentacles of a pink or purplish colour, supported on peduncles of variable lengths, with a cap-like convex head. These tentacles secrete large drops of a viscid secretion (fig. 6).

Besides these tentacles are a number of very minute sessile glands, scarcely visible to the naked eye, colourless, but similar in structure to the tentacles; but with this difference in function, that they do not secrete spontaneously, but must be excited to do so. Both glands and tentacles speedily absorb nitrogenous matter. When an insect alights on a leaf of this fly-catcher, the drops of secretion, with which the tentacles are studded, at once, and readily, adhere to it; and as it moves other drops accumulate, until, at length, bathed with the viscid secretion, it becomes powerless, sinks down and dies, on the small sessile glands with which the leaves are covered. The

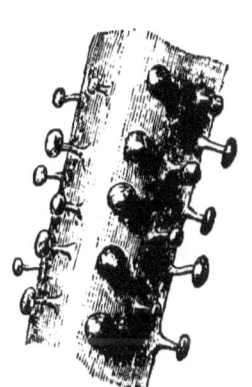

Fig. 6.—Glands on leaf of *Drosophyllum*, magnified.

tentacles have no power of motion, and are not consequently sensitive to the touch. The fly-catching operation is performed by the secretion alone. That the tentacles are capable of absorption is shown by the aggregation of the protoplasm after contact with nitrogenous substances. When the insect falls exhausted and dead, smothered with the viscid secretion of the tentacles, upon the small sessile glands, the contact stimulates the latter to secretion, and it is by their action that the prey is dissolved and assimilated.[1] The process by which the insects are captured differs therefore from that of the Sundews; but after the insect is caught, and deposited upon the small sessile glands, the process of disintegration, and digestion, is evidently the same in all essential particulars.

An allied plant, at the Cape of Good Hope (*Roridula dentata*) probably acts in a similar manner, but no living specimens have been examined. The leaves are studded with glands, which secrete viscid matter, to which insects and other bodies adhere.

The same may be said of an Australian plant, belonging to another genus (*Byblis gigantea*). These can only be named provisionally, as individuals concerning whom further information is desired.

The Sundew family (*Droseraceæ*) includes the six

---

[1] Darwin, "Insectivorous Plants," p. 341.

genera to which we have drawn attention, and of those the true Sundews (*Drosera*) and Venus's Fly-trap (*Dionæa*) are the most important. Of the true Sundews there are no less than one hundred species, "which range in the Old World from the Arctic regions to Southern India, to the Cape of Good Hope, Madagascar, and Australia; and in the New World from Canada to Terra del Fuego." There is every reason to suppose that the same habits, and carnivorous propensities, are common to all, and that, in all this wide range, these humble little bog plants are ever exposing their glittering tentacles to the sun, and luring myriads of insects to their destruction.

> Bright and glorious is that revelation,
>   Written all over this great world of ours;
> Making evident our own creation,
>   In these stars of earth—these golden flowers.

CHAPTER IV.

CARNIVOROUS PLANTS—SIDE-SADDLE FLOWERS.

THE Pitcher-plants, properly so called, are natives of the Old World, their representatives in the New World are called Side-saddle flowers, or Sarracenias. In the true Pitcher-plants the curious pitchers are suspended at the ends of the leaves, of which they are prolongations, but in the Sarracenias the entire leaf is folded and modified into a kind of pitcher. The eight North American species are found in the eastern States, in bogs, and in places covered with shallow water. Their leaves, which give them a character entirely their own, are pitcher-shaped, or rather they are trumpet-shaped, standing erect, collected in tufts, and springing immediately from the ground. They send up at the flowering season one or more slender stems, each of which bears a single flower, which is itself of a peculiar appearance and character, with a fancied resemblance to a side-saddle, and hence the popular name. It has been shown that there are at least two different kinds, or types, of pitcher in this group of plants. In one kind the mouth is open and the lid stands erect, so that the tube receives the rain-water in more or less abundance. In the other kind the mouth of the tube is closed with a lid, and into these the rain can hardly, if ever, find ingress.[1]

As long ago as the year 1815 the fly-catching propensity of these plants was observed and commented upon, in a communication to the President of the Linnæan Society. Many of the assertions then made have since been verified; although at the time they excited but little notice, and perhaps did not receive implicit credence. "If," says the writer, "in the months of May, June, or July, when the leaves of these plants perform their extraordinary functions in the greatest perfection,[2] some of them should be removed to a house and fixed in an erect position, it will soon be perceived that flies are attracted by them. These insects immediately approach the fauces of the leaves, and leaning over their edges appear to sip with eagerness something from their internal surface. In this position they linger, but, at length allured, as it would seem by the pleasures of taste, they enter the tubes. The fly which has thus changed its situation will be seen to stand unsteadily, it totters for a few seconds, slips and falls to the bottom of the

---

[1] " Gardener's Chronicle," August 29th, 1874, p. 260.

[2] These observations relate chiefly to one species, *Sarracenia variolaris*.

tube, where it is either drowned, or attempts in vain to ascend against the points of the hairs. The fly seldom takes wing in its fall and escapes. In a house much infested with flies this entrapment goes on so rapidly that a tube is filled within a few hours, and it becomes necessary to add water, the natural quantity being insufficient to drown the imprisoned insects. The leaves of other species might well be employed as fly-catchers, indeed, I am credibly informed that they are in some neighbourhoods. The leaves of *Sarracenia flava*, although they are very capacious, and often grow to a height of three feet or more, are never found to contain so many insects as those of other species. The cause which attracts flies is evidently a sweet viscid substance resembling honey, secreted by, or exuding from, the internal surface of the tube. From the margin, where it commences, it does not extend lower than one fourth of an inch. The falling of the insect as soon as it enters the tube is wholly attributable to the downward or inverted position of the hairs of the internal surface of the leaf. At the bottom of a tube, split open, the hairs are plainly discernible pointing downwards; as the eye ranges upward they gradually become shorter and attenuated, till at, or just below the surface, covered by the bait, they are no longer perceptible to the naked eye, nor to the most delicate touch. It is here that the fly cannot take a hold sufficiently strong to support itself, but falls. The inability of insects to crawl up against the points of the hairs I have often tested in the most satisfactory manner."[1]

The annexed figure represents the pitchers of the species to which these observations refer (fig. 7). It is also that on which many subsequent and confirmatory experiments were made.

The tissues of the internal, or lining, surfaces of the pitchers in Sarracenia are not identical in all the species. In some, and probably most, there are four kinds of surfaces, proceeding from the mouth downwards to the bottom of the tube. First, there is an attractive surface, often brightly coloured, which occupies the inner face of the lid, and this, in common with the mouth of the pitcher, is covered with minute honey-secreting glands. Then, secondly, there is a conducting surface of glassy cells, which are elongated into conical processes overlapping each other, like the tiles of a house, so as to afford no foothold for an insect attempting to crawl up again. This is succeeded by a large granular surface, which is smooth and polished so as to afford no foothold. And, finally, there is a detentive surface, which occupies the lower part of the pitcher. It is studded

---

[1] Dr. James McBride in "Transactions of the Linnæan Society," vol. xii.

with deflexed rigid hairs, which converge towards the axis of the cavity ; so that an insect, if once amongst them, is effectually detained, and its struggles have no

Fig. 7.— Pitchers of *Sarracenia variolaris*, reduced.

other result than to wedge it lower and more firmly in the pitcher.[1]

---

[1] "Gardener's Chronicle," September 5, 1874, p. 293.

A similar structure in *Sarracenia purpurea* (fig. 8) is thus described by Mr. W. H. Gilburt,[1] in his

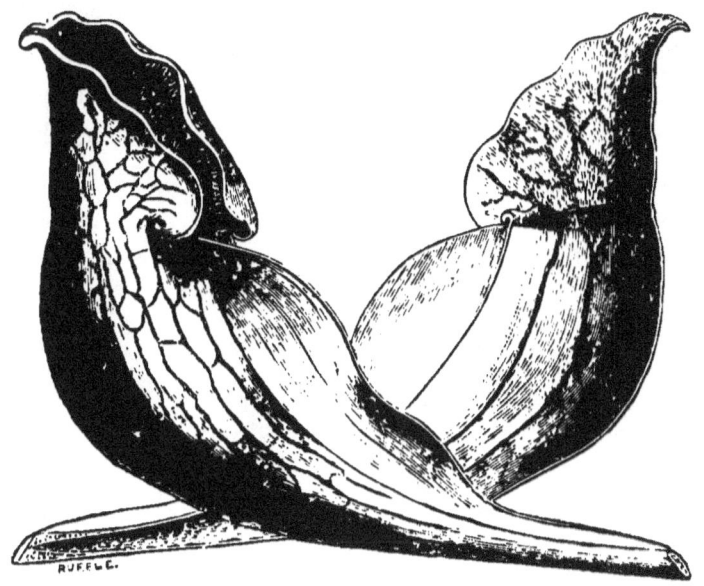

Fig. 8.—Pitcher of *Sarracenia purpurea*, reduced, with section.

memoir on "The Histology of Pitcher-Plants." He says :—" The interior surface of this pitcher is divided

---

[1] W. H. Gilburt in "Journal of the Quekett Microscopical Club," November, 1880, vol. vi., p. 154.

There is a characteristic figure of this Sarracenia in old Gerarde's "Herbal" (1597), where it is called "hollow-leaved sea-lavender," and stated to be copied from Clusius, "for the strangenesse thereof, but hope that some or other that travell into forraine parts may finde this elegant plant, and know it by this small expression, and bring it home with them, that so we may come to a perfecter knowledge thereof."—P. 412.

into four zones. On the first one, or that nearest the mouth of the pitcher, are numerous stomata, and also a large number of strongly developed rigid hairs, which point downward. The second zone is characterised by the fact that each cell of the surface is prolonged downward into a short mammillary process, its wall being striated longitudinally. We next come to a division which is smooth, hairs are entirely absent, and the cells are sinuous in outline. The fourth division is by far the longest, and is crowded with long hairs, the points of which are all directed towards the base, but they are not so stout or strong as those found near the mouth of the pitcher." In explanation the rigid hairs of the upper zone are shown to agree in all respects with an ordinary trichome, being simply the outgrowth of a single cell. These hairs (fig. 9) on their external surface show a few deeply-

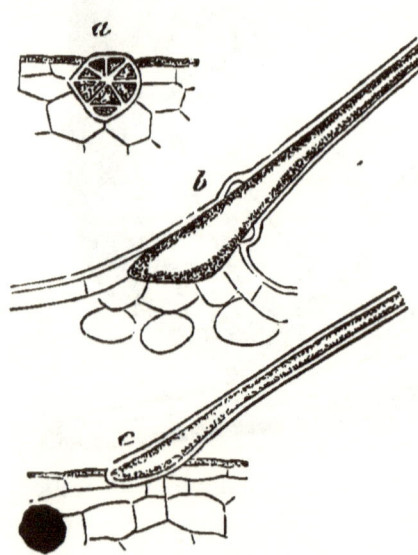

Fig. 9.—*Sarracenia purpurea.*
a Section of gland.
b Hair from upper zone.
c Hair from lower zone.—*Gilburt.*

cut longitudinal striations, in fact, so well marked are they that the hair might almost be described as fluted. Yet are they in error who have described them as made up of a bundle of rod-like cells. Again, he says, " All these modifications of surface are, without doubt, of value to the plant, and in this direction, that while they will allow an insect to enter, and pass down the tube, it is almost impossible for it to return. Thus they become veritable insect-traps. The pitchers of many species contain fluid, but nothing corresponding to a digestive fluid has been detected in them."

As to the fluids, we must carefully guard against misinterpretation. To this end it must be borne in mind that the honey-like, or saccharine, exudation from the lip of the tube, and the fluid, contained in greater or less quantity at the bottom of the tube, are two quite different and distinct substances. The latter will receive attention hereafter, but our present subject is the secretion which is found as a bait or lure at the mouth of the tube. This, combined with the bright colouring, may be fairly assumed to have been provided for some special purpose. Dr. McBride calls it the "cause which attracts flies," and Dr. Millichamp, of South Carolina, set himself to investigate this, and some other disputed points, upon living plants. Having discovered some advanced plants of Sarracenia, he had no difficulty in detecting, in almost every leaf, the sugary secretion or honey-like exudation, noticed by Dr. McBride, and other observers, as being found at the mouth of the tube. "I found it," he writes, "precisely in the place described, save that it extended downwards more than a quarter of an inch, generally half an inch, or even three quarters of an inch. I also found it more sparingly under the arched lid, or upper lip of the leaf, in and among the thick and coarse hairs found there, and which, I believe, are thicker and coarser than those in the lowermost portion of the tube." Dr. McBride, however, failed to trace the continuance of the sugary exudation, which I frequently found glistening, and somewhat viscid, along the whole red or purple-coloured border, or edging of the broad wing, extending from the cleft in the lower lip, *even to the ground*. There is, therefore, a painted or honey-baited pathway, leading directly from the petiole (or the ground itself) up to the mouth, where it extends on each side, as far as the commissures of the lips, from which it runs within, and downwards, for at least half an inch."

"One can now readily understand why ants should so frequently be found among the earliest macerated insects at the base of the tube. Their fondness for saccharine juices is well known, and, while reconnoitring at the base of the leaf, and bent on plunder, they are doubtless soon attracted by the sweets of the honeyed path lying right before them, along which they may eat as they march, until the mouth is reached, where certain destruction awaits them."[1]

In order to determine the character of the saccharine exudation, and whether it possessed any intoxicating properties, Dr. Millichamp collected a large number of mature, and most sugary, leaves, which he placed in vessels of water on reaching home, and sat down before them for two hours watching the result. Flies were soon attracted to the leaves, but by no means greedily, and many were entrapped, the buzzing of unfortunate prisoners being incessant. Finding that he could not see the process with the lids in their normal position, he turned backwards the greater part of the overhanging lid, and let daylight into the prison, so that the whole region of the sugar countries could be seen, and examined, while the flies were busy at their food.

"After turning back the lids of most of the leaves," he says, "the flies would enter as before, a few alighting on the honeyed border of the wing, and walking upward—sipping as they went—to the mouth, and entering at the cleft of the lower lip; others would alight on the top of the lid and then walk under the roof, feeding there; but most, it seemed to me, preferred to alight just at the commissure of the lips,

---

[1] Prof. Asa Gray in " New York Tribune ;" also "Gardener's Chronicle," June 27, 1874.

and either enter the tube immediately there, feeding downward upon the honey pastures, or would linger at the trunk, sipping along the whole edge of the lower lip and eventually enter near the cleft. After entering (which they generally do with great caution and circumspection) they begin again to feed, but their foothold, for some reason or other, seems unsecure, and they occasionally slip, as it appears to me, upon this exquisitely soft and velvety declining pubescence. The nectar is not exuded or smeared over the whole of this surface, but seems disposed in separate little drops. I have seen them regain their foothold after slipping, and continue to sip, but always moving slowly and with apparent caution, as if aware that they are treading on dangerous ground. After sipping their fill they frequently remain motionless, as if satiated with delight, and, in the usual self-congratulatory manner of flies, proceed to rub their legs together, but in reality, I suppose, to cleanse them. It is then they betake themselves to flight, strike themselves against the opposite sides of the prison-house, either upward or downward, generally the former. Obtaining no perch or foothold, they rebound off from this velvety microscopic *chevaux-de-frise*, which lines the inner surface still lower, until, by a series of zigzag but generally downward falling flights, they finally reach the coarser and more bristly pubescence of the lower chamber, where, entangled somewhat, they struggle frantically (but by no means drunk or stupefied), and eventually slide into the pool of death, where, once becoming slimed and saturated with these Lethean waters, they cease from their labours. And even here, although they may cease to struggle, and seem dead, like 'drowned flies,' yet are they only asphyxiated, not by the nectar but by this 'cool and animating fluid, limpid as the morning dew.' After continued asphyxia they die, and after maceration they add to the vigour and sustenance of the plant. And this seems to be the true use of the 'limpid fluid,' for it does not seem to be at all necessary to the killing of the insects (although it does possess that power), the conformation of the funnel of the fly-trap is sufficient to destroy them. They only die the sooner, and the sooner become 'liquid manure.'

"I could never see any indication of unsteadiness or tottering in the sipping flies—nothing save an occasional slip from the uncertain hold which the peculiar pubescence would give, save once or twice while watching intently I saw a fly disappear so quickly downwards that I could not with certainty say whether it was flight or a tumble from stupor or insensibility. But on so many other occasions have I satisfied myself to the contrary, by seeing them fly upward as well as downward, with full vigour of an unhurt unintoxicated insect, that I altogether reject the idea of stupor. I may state that while watching I observed not a single escape when the lid was down, but after I had turned it back on most of the leaves under examination, a few, but only a few, escaped. And those which escaped, after sipping to repletion, seemed in no wise inebriated."[1]

Pursuing these investigations still further, on another occasion he collected the laminæ of about one hundred leaves, all sweet with the exudation. Some of these were placed on a table, after candle-light, and attracted a few hungry flies. They remained many minutes sipping, and would return to sip, seeming to enjoy the evening meal thus afforded them. Of course there could be no entrapment, as only the honey-bearing portions were exposed. The flies ate, and ate, but no unsteadiness, or tottering, or falling, was in a single instance to be seen; and, after having satisfied their appetites, the guests retired for the night. The following day the same tempting viands were placed before the flies, but there was no evidence of a single case of intoxication.

If true that the exudation possessed no intoxicating property, sceptics were next led to inquire how it was that insects were entrapped whilst still in possession of all their instincts and faculties unim-

---

[1] Dr. Asa Gray in "New York Tribune," and "Gardener's Chronicle," June 27, 1874, p. 819.

paired ; and to answer this the same careful observer narrated his experience as to how the flies are entrapped. "The nectar being found below the lower lip for half an inch or more, when the fly is satiated, and makes for flight, he must do so immediately upward for a very short distance, and then somewhat at right angles, to get through the outlet—a rather difficult flight, which perhaps of all insects only a fly might be capable of, but which even he probably is not. This, too, upon the supposition that his head is upward, whereas his head is, I believe, generally downward, or at least parallel with the lip. If in the first position he attempts flight, he is very apt to strike the arch overhead, and, if he escapes that, it is next to an impossibility for him to turn and strike that small space between the projecting (and downward projecting) lid and the lower lip. If with head downward, he is very apt in flight to strike the opposite wall at a still lower angle, and then, from rebound to rebound, get lower and lower until he touches the pool. In almost every instance, therefore, a fly once entering is caught."

The next point for inquiry refers to the fluid contained at the bottom of the tubes or pitchers. What is this fluid which is almost universally present, and what its purpose ? Dr. Millichamp says :—" The first point to decide seemed to be whether the watery fluid found in the leaves was a true secretion of the plant or only rain-water. As I have two or three patches of Sarracenia[1] conveniently near in a neighbouring pine barren, it was no difficult matter to make the necessary examinations. On the 22nd, therefore, the sandy pine-land being very dry and thirsty — no rain having fallen for some days — I visited the plants, which were blooming freely. Many leaves were carefully examined with the throat still closed and impervious to water, and inflated, as they usually are, with air. Upon slight pressure the air would escape, thus opening the throat for inspection. The leaf being tilted, there was almost invariably an escape of fluid—from three to five drops generally—occasionally as many as ten drops, and rarely fifteen drops. It is, therefore, a true secretion, as no rain could possibly have been admitted to the completely-closed and sealed leaf.

"The taste of this secretion was bland, and somewhat mucilaginous, yet seemingly leaving in the mouth a peculiar astringency, recalling very accurately the taste of the root, with which I was quite familiar. So much for the examination of the not yet matured and unopened leaves, in which I may as well remark that I could find no trace of insects, either by puncture, or eggs, or larvæ, nor indeed any *débris* of any kind

---

[1] *Sarracenia variolaris.*

"I next examined a great many perfect leaves with the throat open. In almost every leaf the secretion was to be found, containing generally from ten to fifteen drops, very rarely a half drachm. Even in these open leaves the admission of rain-water is next to impossible, so completely does the upper lid overhang the mouth or throat, like the projecting eaves of a house. Unless in a severe rain-storm, and perhaps not even then, would this be possible.

"With very rare exceptions dead and decaying, or, more properly, macerated insects were to be found packed at the base of the tube—most frequently a large red ant—also beetles, bugs, flies, &c., and invariably within the decaying mass one or more small white worms, perhaps the larvæ of insects hatched within the putrefying mass.[1]"

Accepting the explanation as satisfactory, we arrive at the conclusion that the tubes or pitchers of the Sarracenias have the power of secreting a limpid fluid, in addition to the honey-like secretion of the lip, and that this fluid is collected at the bottom of the tube, into which the captured insects fall. Having sipped the nectar which was spread at the mouth, a harmless but seductive lure, spread upon treacherous ground, we have seen that the structure of the tube was favourable to the retention of the insects, and

---

[1] Dr. Asa Gray in "New York Tribune," 1874, and "Gardener's Chronicle," June 27, 1874, p. 818.

that numbers of them gradually, but surely, found their way to the bottom. Here another secretion is stored, which doubtless possesses some properties of its own, and has some function to perform. In order to determine this, Dr. Millichamp proceeded to the investigation of the fluid secretion of the pitchers.

"By draining every leaf plucked of its few drops of juice, I collected about half an ounce of the secretion in a vial, with which I made careful experiments in testing its intoxicating effect upon insects. My subjects were chiefly house-flies. About half a drachm to a drachm of the secretion was placed in a small receptacle, and the flies thrown in from time to time, the liquor not being deep enough to immerse them completely, but enabling them to walk about in it without swimming and the risk of being drowned. Some twenty flies were experimented with. At first the fly makes an effort to escape, though apparently he never uses his wings in doing so—the fluid, though not seemingly very tenacious, seems quickly to saturate them, and so clings to them, and clogs them, as to render flight impossible. A fly, when thrown in water, is very apt to escape, as the fluid seems to run from its wings, but none of these escaped from the bath of the Sarracenia secretion. In their efforts to escape they soon get unsteady in their movements, and tumble sometimes on their backs; recovering, they make more active and frantic efforts, but, very quickly, stupor seems to overtake them, and they then turn upon their sides either dead (as I at first supposed) or in profound anæsthesia.

"I had no doubt, from the complete cessation of all motion, and from their soaked and saturated condition, that they were dead, and, like dead men they were 'laid out,' from time to time, as they succumbed to the powerful liquor; but to my great surprise, after a longer or shorter interval—from a half-hour to an hour or more—they indicated signs of returning life by slight motions of the legs and wings, or body. Their recovery was very gradual, and eventually, when they crawled away, they seemed badly crippled and worsted by their truly Circean bath. After contact with the secretion, the flies which were first thrown in became still, seemingly dead, in about half a minute, but whether from exposure to the air, or exhausted by action on these insects, the liquor did not seem to be so intoxicating with those last exposed to its influence. Anæsthesia or intoxication certainly did not occur so quickly; it took from three to five minutes generally, and in one rebellious subject it took at least ten minutes for him to receive his *coup-de-grâce*. A cockroach thrown in succumbed almost immediately, as did also a small moth, and much more slowly a common house-spider. On the recovery of the latter it was almost painful to witness his unsteady motions, and seeing him dragging his slow length along.

"Without doubt, therefore, the secretion found in the tubes of Sarracenia is intoxicating, or anæsthetic, or narcotic, or by whatever word you may prefer to indicate that condition to which these insects succumb. I forgot to mention that while experimenting as above I also threw several flies in water—a few escaped, one remained for some hours, still 'paddling' and undrowned. A large 'blue fly' was also repeatedly immersed in a weak solution of gum-arabic (in imitation of the fluid of Sarracenia) but he remained unhurt all night, and I liberated him in the morning."[1]

Having thus far recapitulated the results of most of the experiments which have been undertaken in connexion with the Sarracenias, it is necessary to see how matters stand with them in relationship to their insectivorous proclivities. They possess pitchers, or receptacles capable of holding and retaining insects. These receptacles are furnished with glands which excrete around the mouth of the pitcher saccharine juice peculiarly attractive to insects, although of no special service to the plant. Insects seduced by this nectar congregate around the mouths of the pitchers, the sides of which are so constructed that they present

---

[1] Prof. Asa Gray in "New York Tribune," 1874, and "Gardener's Chronicle," June 27, 1874, p. 818.

no obstacles to downward precipitation. The inner lining of these pitchers consists of four zones, the lowest present hairs, in considerable numbers, pointing downwards, so that insects which have once fallen down are unable to get out again, in fact, they become veritable traps. Fluid accumulates at the bottom of these vessels, in which imprisoned insects are drowned, whether or not it is intoxicating in its properties matters but little. The whole structure and adaptation is that of a "fly-catcher." Numerous insects have constantly been found at the bottom of the pitchers. Thus far, although the evidence is circumstantial, it seems to indicate insectivorous propensities, but no experiments yet instituted have demonstrated the presence of an acid secretion, which should aid in the digestion of the captured insects. That the plant has the power of catching insects will not be denied, but as yet there is no proof that it has the power of digesting them. As Mr. W. H. Gilburt has said, "The pitchers contain fluid, but nothing corresponding to a digestive fluid has been detected in them; so that if the insects which perish in the pitcher are of any value to the plant, and afford any nutriment, it must be simply by maceration, and the glands can be regarded as absorbent only."[1] Unfortunately, however, the evidence of absorbent glands is faulty.

---

[1] "Journal of Quekett Microscopical Club," vol. vi., p. 157.

We have met with no indications in any of the details of experiments, of aggregation of protoplasm, which, as we have seen elsewhere, accompanies the absorption of animal matter. Hence, for the present, the admission of the Sarracenias into the circle of insectivorous plants can only be tentative. There may be strong presumption that it would not have become a fly-catcher, on such an extensive scale, if such a proceeding were not in some way beneficial to the plant. That, unless insects are of some service, it may be regarded as a waste of power, to secrete a honeyed juice around the open jaws of death, and lure the unoffending flies to certain destruction, in wanton mischief. The fact, perhaps, would be better stated in such terms as would indicate, that hitherto the observations are incomplete, and, although strongly suspected, the Sarracenias can only be charged with destroying insects, and not with devouring them.

It is manifestly difficult to suggest any plausible theory to account for the presence of insects, nay, more, the special adaptation for the capture of insects, if we reject the carnivorous hypothesis. We could not accept the suggestion that they simply store up insects for certain insectivorous birds, nor that insects are collected in their receptacles in order to furnish food to some larvæ, which are developed from the eggs of other insects, which are deposited there by choice or chance. The only feasible theory would be one which can associate the decaying insects in some manner with the roots, by demonstration that the nitrogenous matter is conveyed from the pitchers by decay or puncture at the base, or some such means, whereby the generous fluid may be conveyed into the soil, and absorbed by the plant in the ordinary manner.

"It must be quite certain that the insects which go on accumulating in the pitchers of Sarracenia are far in excess of its needs for any legitimate process of digestion. They decompose, and various insects, too wary to be entrapped themselves, seem habitually to drop their eggs into the open mouth of the pitchers, to take advantage of the accumulation of food. The old pitchers are consequently found to contain living larvæ and maggots, a sufficient proof that the original properties of the fluid which they secreted must have become exhausted. And Barton says that various insectivorous birds slit open the pitchers with their beaks to get at their contents."

At one of the meetings of the Scientific Committee of the Royal Horticultural Society, during 1879, it was reported that a cultivator of Sarracenias in this country had one species, the pitchers of which furnished such an attraction to flies, that they soon became completely gorged; and thus, as the pitchers were destroyed, the plants could not be successfully cultivated until a device was discovered, which consisted in blocking the mouths of the pitchers with cotton-wool, which had the desired effect, and afterwards the cultivation of that species proceeded satisfactorily. This fact would certainly indicate that animal food is not essential to at least one species.

In an interesting communication on carnivorous plants, from an entomological point of view, Professor C. V. Riley (American Department of Agriculture) has described two insects, and given details of their life-history, which live in, and are parasitic upon, the contents of the pitchers of Sarracenia. He has shown that these insects flourish upon, the ruin of the many victims of the honeyed lure which these pitchers present, and as a summary of his observations he concludes thus :—

"1. There is no reason to doubt, but every reason to believe, that Sarracenia is a truly insectivorous plant, and that by its secretions and structure it is eminently fitted to capture its prey.

"2. That those insects most easily digested (if I may use the term), and most useful to the plant, are principally ants and small flies, which are lured to their graves by the honeyed paths; and that most of the larger insects, which are not attracted by sweets, get in by accident and fall victims to the peculiar mechanical structure of the pitcher.

"3. That the only benefit to the plant is from the

liquid manure resulting from the putrescent captured insects.

"4. That the parasitic moth is a mere intruder, its larva sharing the food obtained by the plant.

"5. That the fly (which also breeds in the pitchers) has no other connexion with the plant than that of a destroyer, though its greatest injury is done after the leaf has performed its most important functions. Almost every plant has its peculiar insect enemy, and Sarracenia, with all its dangers to insect life generally, is no exception to the rule."[1]

Another plant,[2] very similar to the Sarracenias, is found at an elevation of 5,000 feet on the Sierra Nevada of California. "It has pitchers of two forms; one, peculiar to the infant state of the plant, consists of narrow, somewhat twisted trumpet-shaped tubes, with very oblique open mouths, the dorsal lip of which is drawn out into a long, slender, arching, scarlet hood that hardly closes the mouth. The slight twist in the tube causes these mouths to point in various directions, and they entrap very small insects only. Before arriving at a state of maturity the plant bears much larger, nearly erect pitchers, also twisted, with the lip produced into a large inflated hood that completely arches over a very small entrance to the cavity of the pitcher. A singular

---

[1] "Science-Gossip," 1874, pp. 273, 5. [2] *Darlingtonia.*

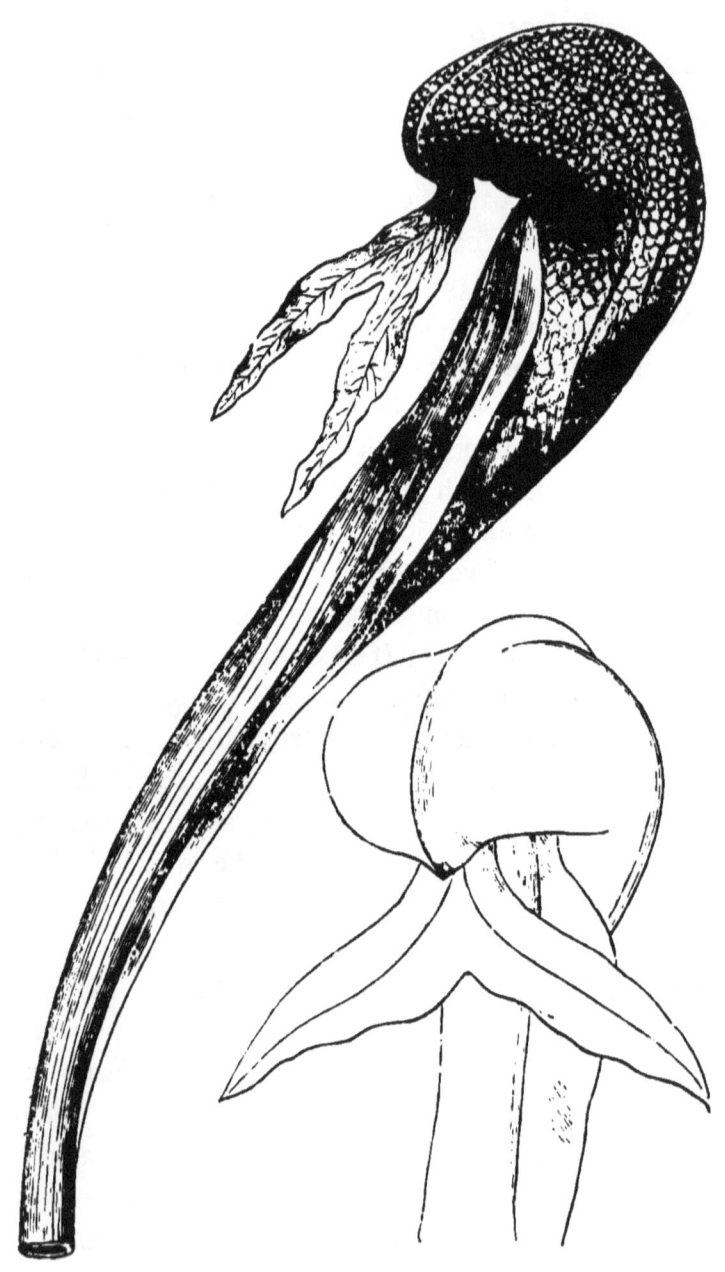

Fig. 10.—Pitchers of *Darlingtonia*.

orange-red, flabby, two-lobed organ hangs from the end of the hood, right in front of the entrance, which, according to Professor Asa Gray, is smeared with honey on its inner surface. These pitchers are crammed with large insects, especially moths, which decompose in them, and result in a putrid mass."[1] It is a curious fact that the change from the slender, opened-mouthed, to the inflated, close-mouthed pitchers, is absolutely sudden in each individual plant. No intermediate stages are to be found, so that the young pitchers almost represent those Sarracenias which have open mouths and erect lids, and the mature pitchers represent those other Sarracenias which have closed mouths and globose lids.

The minute structure of the pitchers corresponds in some respects, but differs in others from that of Sarracenia. According to Mr. Gilburt, the inner surface of the pitcher is divided into two zones: the upper one is furnished with short, thick, spike-like trichomes, comparatively wide apart, while the lower one has the same kind and arrangement as exists in Sarracenia. " The glands are the simplest in structure of any found in the group. The epidermal cells of the upper zone of the pitchers have the common sinuous line, but scattered among them are a considerable number of large spherical cells, one portion

---

[1] " Gardener's Chronicle," August 29, 1874, p. 260.

of their wall being exposed at the surface, and the remainder dipping below the epidermis into the subjacent tissue. These glands are very inconspicuous when the tissue is in its natural condition, but if the colour is discharged from a portion of the plant by means of alcohol they are at once apparent, and their contents are seen to be different from that of the surrounding cells, the chlorophyll corpuscles being absent. These I take to be glands, but what their function may be, if any, is rather difficult to imagine."[1] We have given these details thus minutely because everything connected with the structure of these plants is of interest so long as the mystery is unexplained wherefore they catch insects. The Darlingtonia, as the Sarracenia, gives no indication of possessing the faculty of digestion, and the remarks we have made on the latter will apply equally to the former. Whatever the future verdict may be, whether guilty or not guilty of being carnivorous, they still would claim a place in this work for the singularity of their appearance, the extraordinary form of their flowers, their peculiar trumpet-shaped receptacles, their fly-catching arrangements, and the mystery which enshrouds their domestic economy.

---

[1] W. H. Gilburt, "Journal of Quekett Microscopical Club," vi., p. 158.

CHAPTER V.

CARNIVOROUS PLANTS—PITCHER-PLANTS.

THERE are some plants which have commended themselves to notice either by their singular form, peculiar habit, showy flowers, or beautiful odour. Before carnivorous plants attracted any attention on account of their flesh-devouring proclivities, the Pitcher-plants had acquired notoriety, not on account of their showy flowers or beautiful odours—because these are attractions which they do not possess—but simply on account of their singular form. The pitchers, from whence the name is derived, hang suspended at the ends of the leaves, of which they are simply prolongations and modifications. Most Pitcher-plants consist of a clump of long, narrow green leaves. The extremities of the latter are attenuated down to the midrib, which becomes reduced to a cord, at the end of which hang suspended, one from each of many of the leaves, a curious bag or pouch, not unlike a small and delicate jug or pitcher, with a smaller leaf-like flap hanging over the mouth like a lid. These pitchers usually contain a little fluid, looking like water at the bottom, in which are drowned insects. Such were the Pitcher-plants to our forefathers, and they were regarded simply as "curiosities of vegetation." To us they are something more, now that their history is better known, and for reasons which it shall be our object to explain.

Botanically, the Pitcher-plants proper are known by the name of Nepenthes, an old classical name, the application of which to these plants is somewhat obscure. One writer has attempted an apology for it in the following manner:—"I have often wondered why Linnæus gave to this genus the name of Nepenthes. Every reader of classic story remembers that when Telemachus reached the court of Menelaus, tired and famished, the beautiful Helen gave him nepenthe to drink. No one has ever been able to say what this nepenthe was, though no doubt one of the 'drowsy syrups of the East.' Johnson defines nepenthe as an 'herb that drives away sadness.' Linnæus, perhaps, intended to refer to the tankard-like structure, so like also in the original species to a hot-water jug with its lid. Sometimes I am disposed to think that old Homer may have meant by nepenthe no physical beverage, but the sweet graces of Helen's queenly and consummate hospitality, and welcome, touching, as they did, her guest's inmost feelings of love and reverence. If so, Nepenthes is well applied to its present owner, for assuredly no plant appeals more strongly to our sense of the admirable and the unique."[1]

These tropical plants can only be cultivated in hot-houses in this country, and hence there are many persons to whom they are utter strangers. It may be true that all recent horticultural exhibitions have included specimens, but there are thousands of unfortunate individuals who can never visit "flower-shows," although there are but few in the neighbourhood of the metropolis who could not search out the Pitcher-plant in that favourite holiday resort—Kew Gardens. Travellers have described for us the appearance of these plants in their native homes, and especially those who have visited Borneo and the other islands of the Indian archipelago. Amongst others, Mr. Alfred Wallace thus alludes to them. He says:—"We had been told we should find water at Padangbatu, but we looked about for it in vain, as we were exceedingly thirsty. At last we turned to the Pitcher-plants, but the water contained in the pitchers (about half a pint in each) was full of insects, and otherwise uninviting. On tasting it, however, we found it very palatable, though rather warm, and we all quenched our thirst from these natural jugs."[2] And again, when at Borneo, the same traveller writes:—"The wonderful Pitcher-

---

[1] "Gardener's Chronicle," January 9, 1875.
[2] Wallace, "Malay Archipelago," vol. i., pp. 49 and 127.

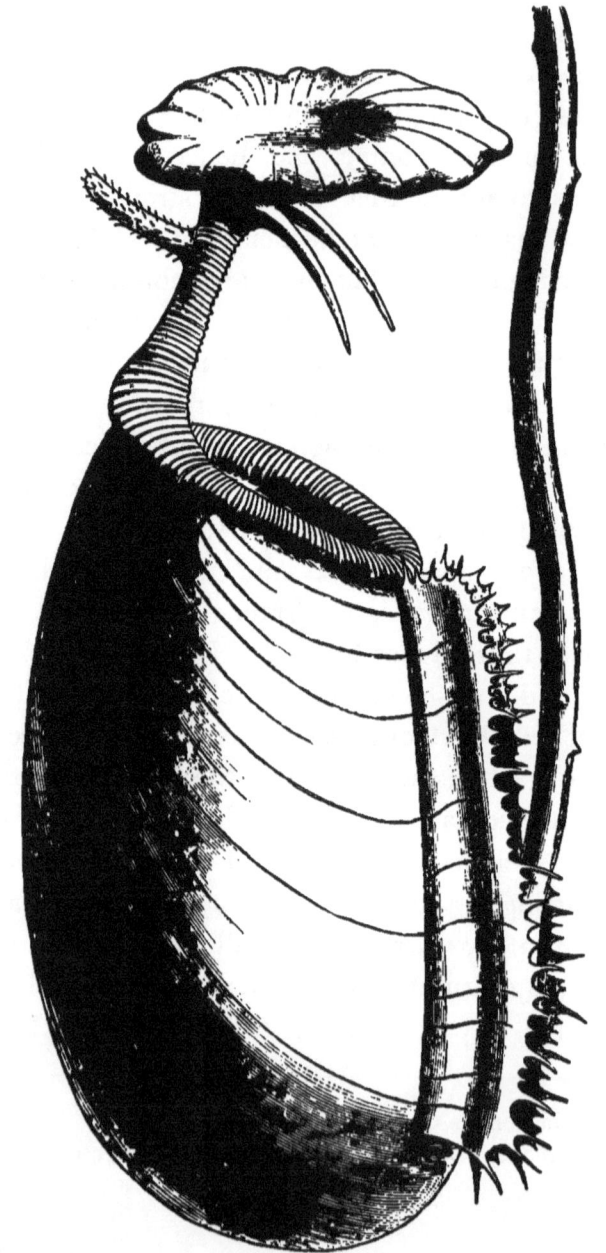

Fig. 11.—Pitcher of *Nepenthes bicalcarata*.

plants, forming the genus Nepenthes of botanists, here reach their greatest development. Every mountain-top abounds with them, running along the ground or climbing over shrubs and stunted trees; their elegant pitchers hanging in every direction. Some of these are long and slender, resembling in form the beautiful Philippine lace-sponge, which has now become so common; others are broad and short; their colours are green, variously tinted, and mottled with red or purple. The finest yet known were obtained on the summit of Kini-balou, in north-west Borneo. One of the broad sort[1] will hold two quarts of water in its pitcher. Another[2] has a narrow pitcher 20 inches long, while the plant itself grows to the length of 20 feet." In 1847, when Lindley published the second edition of his "Vegetable Kingdom," he recorded, with somewhat of doubt, the number of different species as six, whereas, so many have been discovered since, that we may consider them equal to five times that number.

There are, says Dr. Hooker, "upwards of thirty species of Nepenthes, natives of the hotter parts of the Asiatic archipelago, from Borneo to Ceylon, with a few outlying species in New Caledonia, in tropical Australia, and in the Seychelles Islands on the African coast. The pitchers are abundantly pro-

---

[1] *Nepenthes rajah.*   [2] *Nepenthes Edwardsiana.*

duced, especially during the younger state of the plants. They present very considerable modifications of form and external structure, and vary greatly in size, from little more than an inch to almost a foot in length; one species indeed, from the mountains of Borneo, has pitchers which, including the lid, measure a foot and a half, and the capacious bowl is large enough to drown a small animal or bird."

"In most species the pitchers are of two forms, one pertaining to the young, the other to the old state of the plant, the transition from one form to the other being gradual. Those of the young state are shorter and more inflated; they have broad fringed longitudinal wings on the outside, which are probably guides to lead insects to the mouth; the lid is smaller and more open, and the whole interior surface is covered with secreting glands. Being formed near the root of the plant, these pitchers often rest on the ground, and in species which do not form leaves near the root they are sometimes suspended from stalks which may be fully a yard long, and which bring them to the ground. In the older state of the plant the pitchers are usually much longer, narrower, and less inflated, trumpet-shaped; the wings also are narrower, less fringed, or almost absent. The lid is larger and slants over the mouth, and only the lower part of the pitcher is covered with secreting glands, the upper part presenting a tissue of different character. The difference of structure in these two forms of pitcher, considered in reference to their different positions on the plant, forces the conclusion on the mind that the one form is intended for ground game, the other for winged game. In all cases the mouth of the pitcher is furnished with a thickened corrugated rim, which serves three purposes : it strengthens the mouth, and keeps it distended ; it secretes honey, and it is in various species developed into a funnel-shaped tube, that descends into the pitcher, and prevents the escape of insects, or into a row of incurved hooks, that are in some cases strong enough to retain a small bird, should it, when in search of water or of insects, thrust its body beyond a certain length into the pitcher. In one species (*Nepenthes bicalcarata*), there are also two strong pointed hooks, or teeth, which are directed downwards towards the mouth of the pitcher. Such appendages would doubtless be of service in preventing the free exit of any large insect after it had once entered the pitcher (*see* fig. 11).

The attractive surfaces of Nepenthes are two, those namely of the rim of the pitcher, and of the under surface of the lid, which is provided in almost every species with honey-secreting glands, often in great abundance. It is a singular fact that the only species known to the writer of these observations, in which the honey glands on the lid were absent, was a species[1] in which the lid, unlike that of other species, is thrown back horizontally. The secretion of honey on a lid so placed would tend to lure insects away from the pitcher instead of into it.

From the mouth downwards, for a variable distance inside the pitchers, the glassy glaucous surface affords no foothold for insects. The rest is entirely occupied with the secretive surface, which consists of a cellular floor crowded with spherical glands in inconceivable numbers. Each gland resembles the honey-glands of the lid, semicircular, with the mouth downwards, so that the secretive fluid all falls to the bottom of the pitcher. In one species[2] three thousand of these glands were ascertained by Dr. Hooker to occur on a square inch of the inner surface of the pitcher, and upwards of a million in an ordinary-sized pitcher. The glands secrete the fluid which is contained at the bottom of the pitchers previous to their opening, and this fluid is always acid. When the fluid is emptied out of a fully-formed pitcher, that has not received animal matter, it forms again, but in comparatively very small quantities, and the formation goes on for many days, even after the pitcher has been removed from the plant. "I do not find," says Dr. Hooker, "that placing inorganic substances in

---

[1] *Nepenthes ampullacea.*   [2] *Nepenthes Rafflesiana.*

the fluid causes an increased secretion, but I have twice observed a considerable increase of fluid in pitchers after putting animal matter in the fluid."[1]

A series of experiments performed with the pitchers of these Pitcher-plants, resembled those applied previously to the Sundews and Fly-trap, with similar results. White of egg, raw meat, fibrine, and cartilage were employed for feeding. In all cases the action was most evident, and in some surprising. After twenty-four hours' immersion, the edges of the cubes of white of egg were eaten away, and the surfaces gelatinised. Fragments of meat were rapidly reduced, and pieces of fibrine weighing several grains were dissolved, and had totally disappeared in two or three days. With cartilage the action was most remarkable. Lumps of this, weighing eight and ten grains, were half-gelatinised in twenty-four hours, and in three days the whole mass was greatly diminished, and reduced to a clear, transparent jelly.

That this action, which is comparable to digestion, is not wholly due to the secretion, as at first deposited, seems probable, since very little change took place in any of the substances when placed in the fluid drawn from the pitchers, and put in glass tubes, nor even in substances immersed in the pitchers, when the plants have been removed into a room the temperature of

---

[1] See "Gardener's Chronicle," September 5, 1874, p. 293.

which was far below that of the normal temperature in which the plant flourishes. In the latter case, as soon as the plant was taken back into a higher, and more normal temperature, the immersed substances were immediately acted upon.

Comparing the action of meat, and other substances, placed in tubes containing fluid drawn from the pitchers, with similar substances placed in tubes containing distilled water, it was observed that disintegration was three times as rapid in the fluid as in the water, but this disintegration was wholly different in its character from that which followed the immersion of like substances in the fluid still contained in the pitchers of growing plants.

The conclusions arrived at from the above experiments were thus summarised. " From these observations it would appear probable that a substance acting as pepsine is given off from the inner wall of the pitcher, but chiefly after placing animal matter in the acid fluid ; but whether this active agent flows from the glands, or from the cellular tissue in which they are imbedded, I have no evidence to show."[1]

Reverting to the lids of the pitchers, on which honey-secreting glands are stated to have been found in all species but one, it must be recorded that Dr.

---

[1] Dr. Hooker's Address at British Association, Belfast Meeting.

Lawson Tait disputes the fact. He says: "On the under surface of the lids of most pitchers are to be seen glands identical in structure with those occurring on the inner surface of the pitcher. Dr. Hooker believes these to be honey-glands, but I differ from that eminent authority, for the following reasons:— That they are identical in structure with the true digestive glands, and that they are better marked in the pitchers where the lids cover the mouth completely, than in those which do not; that in many (such as in *N. distillatoria*) they are hooded in exactly the same way as the glands of the pitcher; that when the gland is excited by food, I have been able to detect acid secretion collected in the hoods of the lid-glands of *N. distillatoria;* that nectaries are usually very inconspicuous, and only a small spot of tissue which, without being transformed, produces the nectar."[1]

The above appears to be very circumstantial, but perhaps it did not occur to the writer to inquire for what purpose acid-secreting, or digestive, glands occur on the lid of the pitcher, or whether the acid secretion found collected in the hoods might not have been accidental, or whether he might not have overlooked the "inconspicuous" honey-secreting glands,

---

[1] Structure of Pitcher-plants, in "Midland Naturalist," March, 1880.

and whether himself, and the authority with whom he joins issue, may not, after all, have seen and

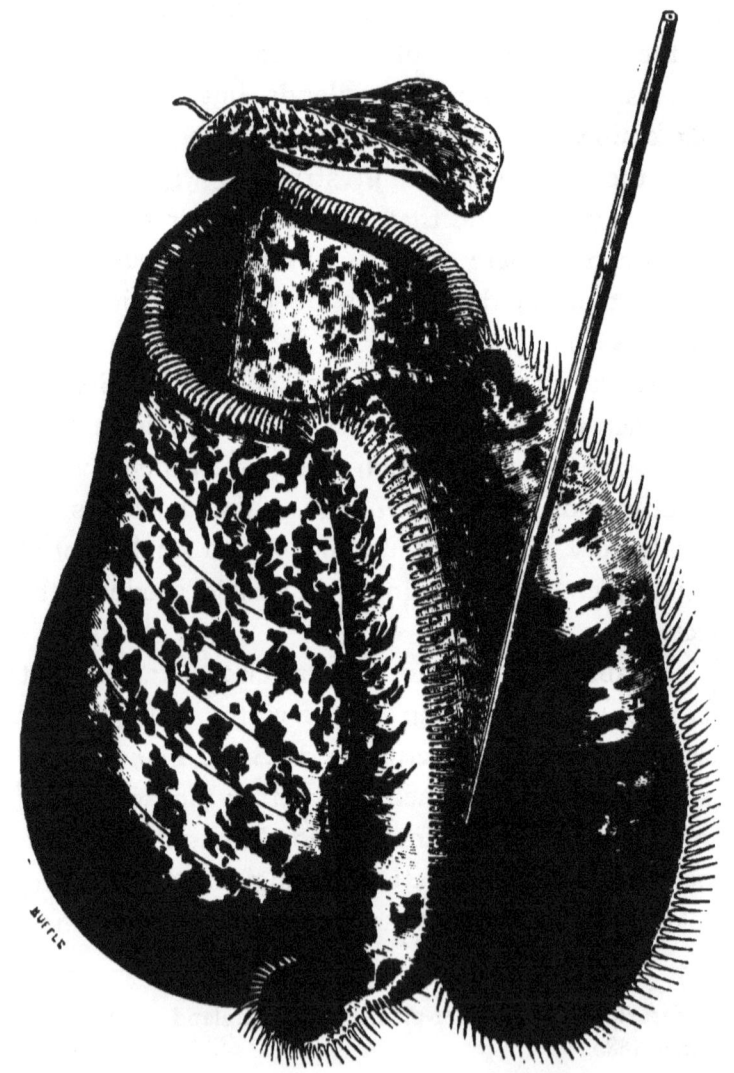

Fig. 12.—Pitcher of *Nepenthes Chelsoni*.

examined two different classes of glands. Had they both really seen the *same* glands, it is difficult to conceive how they could have arrived at such opposite conclusions. From analogy, as well as probability, the presumption is certainly most strongly in favour of glands at the orifice secreting a sweet, attractive, rather than a digestive fluid.

The form of the glands in the pitchers has been detailed by Mr. Gilburt, as seen in one species, and this applies pretty generally to all. "All the glands we have hitherto treated of," he writes, " are imbedded; these project entirely above the surface. They are also over-arched by a canopy crescentic in form, and have each of them direct communication with a twig of the vascular system. The arrangement of the position of these glands is quite irregular and unsymmetrical. The canopies, or hoods, however, of those nearest the mouth of the pitcher cover the glands more completely than do those lower down, while at the base the glands are wholly exposed. The figure shows the relation of hood to gland about midway between the two extremes."

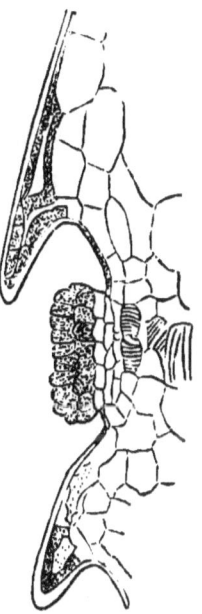

Fig. 13.—Section of hood of *Nepenthes Chelsoni* (*Gilburt*).

"The glands vary very much in size, and are composed of five or six layers of thin-walled cells, without intercellular spaces. The cells of the superficial layer have their longest diameter at right angles to the surface, those of the next layer are about equal in all directions, while those beneath are flattened in all horizontally, and have walls of extreme thinness. At the base of the gland are generally to be seen one or two spirally-marked cells, in a group of which the vascular twig always terminates."[1]

Dr. Lawson Tait constructed an elaborate table, in which are shown the variations in size of the glands found in the pitchers of one species. Progressing downwards, from the opening to the bottom of the pitcher, the glands were measured at every five millimetres, commencing in size at ·045, and gradually increasing until they attained a diameter of ·2 mm. Through the whole course downwards an increase was maintained, which was gradual for the first third of the distance, then suddenly the diameter is almost doubled. During the second third of the course the diameter is increased about one-third, whilst in the last third the increase is very slight. The estimated number of glands in a square millimetre are 73 in the upper zone, 25 in the middle

---

[1] W. H. Gilburt in "Journal of Quekett Microscopical Club," vi., p. 162.

zone, and 36 in the lowest. So that, where the diameter of the glands only ranges between ·045 and ·07 mm., the number is 73 in a given space (one square millimetre); but when the glands are ·135 to ·19 mm. in diameter, or twice to nearly three times as large, there are only one-third of the number, or 25, in the same area.

In the upper zone (of *Nepenthes Rafflesiana*), where the glands are small, though numerous, they are wholly covered by the hood. " The complete covering of the glands in this zone," writes Dr. Lawson Tait,[1] " may be of advantage in protecting them and their secretion from accident and the depredation of insects, for the glands here are much more likely to be uncovered by water than those further down. I think it is also very likely that these hoods store up the digestive principles of the pitcher until they are required, or until it is washed out by the contact of water, it being retained in their cavities by capillary attraction."

"In the second zone the glands gradually alter from a round shape to an oval one, increasing at the same time in size as they are viewed from above downwards, and they become less covered by the hoods. The relative gland area is also greatly increased. The greatest amount of work would necessarily fall

---

[1] " Midland Naturalist," March, 1880, p. 62.

on the third zone, so we have here the glands at their maximum, and almost uncovered by the hoods, which still remain in existence however. The glands are so large and so close that the bulk of the surface is occupied by them."

If we were to summarise these details we should arrive at something like the following conclusions as to the pitchers of Nepenthes. That they are adapted for the capture and retention of insects; that at the orifice are certain attractions, such as the production of a sweet fluid, which would be likely to allure insects into the traps; that the mouth is protected by an overhanging lid, which would prevent the falling in of small and useless objects, but insufficient to obstruct the entrance of living prey; that this lid may also prevent the admission of an excess of external moisture; that the internal structure is extensively glandular, the glands being elevated, but at the same time protected by a hood; that the glands are largest, cover the largest surface, and are least protected at the bottom and the lower third of the inner surface; that these glands secrete a digestive fluid, with an acid reaction; that insects are commonly found at the bottom of the pitchers where they become disintegrated; that they are probably digested by the excreted fluid, and their soluble nitrogenous matter absorbed and assimilated for the advantage of the plant. Hitherto we have failed to discover, in the records of the investigations, any satisfactory evidence of the aggregation of protoplasm, which in other plants has afforded so strong an argument in favour of the absorption of animal matter. From all this, if the summary is a fair one, we are naturally led to conclude that the pitchers of the pitcher-plants are traps to catch animals, as well as stomachs to digest them, and that there are sufficient grounds for including the species of Nepenthes amongst insectivorous or carnivorous plants. Further investigation may probably discover something analogous to aggregation.

Fig. 14.—Pitcher of *Cephalotus*.

The Australian pitcher-plant (*Cephalotus follicularis*, fig. 15) is a smaller and much more unpretending plant than the true pitcher plants of the tropics. The leaves are produced in a rosette close to the ground, and consist of true leaves and pitchers. The latter are attached to their footstalk in an opposite direction to the pitchers of Nepenthes, the mouth being directed towards the axis. The minute structure and anatomy of these pitchers has been pretty exhaustively examined.[1] The thimble-shaped pitchers are surmounted by a lid, which is elevated, so that the entrance to the pitchers is open and free. We are not concerned with the external structure, which is fully detailed in the memoir by Professor Dickson.

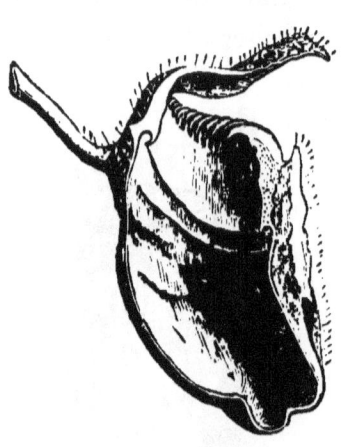

Fig. 15.— Section of pitcher of Cephalotus.

The mouth of the pitcher is furnished with a corrugated rim which ends abruptly on the inner margin in a row of inflexed teeth, extending along the front of the orifice to the base of the lid (see section, fig. 15). Below the rim is a ledge extending round the inside of the pitcher, with its acute edge projecting downwards into the cavity, forming a kind of contracted neck. This is called the conducting-shelf. Below this, again, the upper two-thirds of the walls are smooth and glandular. At the lower margin of this smooth surface an oblique curved elevation

---

[1] By Professor Dickson in the "Journal of Botany," for January, 1878, vol. vii., p. 1. Confirmed by Mr. W. H. Gilburt in "Journal of Quekett Microscopical Club," November, 1880, vol. vi., p. 159; and by Dr. Lawson Tait in "Midland Naturalist."

extends on each side, and below all is the bottom of the pitcher, which is smooth and without glands. The surface of the conducting-shelf is furnished with hairs projecting downwards.

Looking at this arrangement as that of a fly-catching apparatus, it seems to be admirably adapted to the purpose, the projecting ledge, the downward directed hairs with which it is furnished, and the incurved teeth of the rim, all present obstacles to the exit of any insect which may enter.

Small glands with six cells, two of which are central and four peripheral, are scattered over the under-surface of the lid and upon the rim. The upper walls of the interior immediately beneath the ledge contain numerous large glands of from thirty to forty cells.

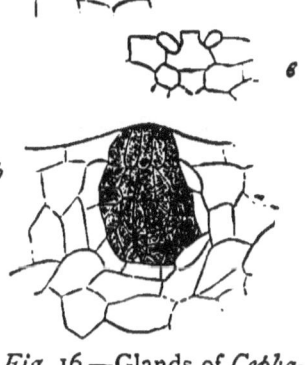

Fig. 16.—Glands of *Cephalotus*, in section (Gilburt).

The oblique curved elevations below the glandular walls are studded with similar glands, but nearly twice as large. Mixed with these are smaller hyaline glands, which consist of a central oval cell surrounded by two to four others; the appearance is somewhat that of stomata, but the central cell, as Mr. Gilburt believes, is glandular, and

Dr. Dickson surmises that its secretion may serve to dilute the other secreted matter (fig. 16, *c*).

Of the oblique curved elevations on which these latter glands are situated, Mr. Gilburt writes: " On the lateral walls of the pitchers near the base there are two crescent-shaped raised patches inclined forward and downward towards the front of the pitcher at a very acute angle. These are without doubt the most active secreting organs of the plant. In mature pitchers these patches are coloured a deep crimson; whether they serve in any sense as a bait to lure the insects into the pitcher I do not know, but such an idea is not improbable." And Dr. Dickson says that "the red coloured cell-contents very soon after injury of the cells, or treatment by reagents, change to a bright blue."

Insects are undoubtedly caught in these pitchers, and their destiny is summed up in a few words by Dr. Tait: " In two pitchers I found insects bathed in fluid with a strongly acid reaction, and this fluid digested shreds of albumen exactly as I found the fluid of Nepenthes pitchers did. I conclude, therefore, that a true digestion of its victims is carried on by the Cephalotus pitchers."

The reader will experience no difficulty in coming to conclusions upon the facts we have submitted. This chapter may conclude with a reference to two plants mentioned by Knapp, but which have not received special attention since the subject of carnivorous plants has been fairly under investigation. "The different manner," he says, "in which vegetables exert their organic powers to effect the destruction of insects is not, perhaps, unworthy of a brief notice: some accomplish it by means of elastic or irritable actions, adhesive substances, and so forth; but we have another plant in our green-houses, the glaucous birthwort (*Aristolochia glauca*), that effects these purposes without any of these means, but principally by conformation. The whole internal surface of the tubular flower is beset with minute strong spines, pointing downwards; these present no impediment to the descent of the animal which may seek for the sweet liquor lodged upon the nectary at the base of the blossom, nor is there any obstruction provided for its return by means of valves or contractions, the tube remaining open; but the creature cannot crawl up by reason of the inverted spines; and, to prevent its escape by flying up the tube, the flower makes an extraordinary curve, bending up like a horn, so that any winged creature must be beaten back by striking against the roof of this neck as often as it attempts to mount, and, falling back to the bulbous prison at the base of the flower, dies by confinement and starvation, and there we find it: a certain number of these perishing, the blossom fades and drops off."

Of the other plant he writes: "It is a perplexing matter to reconcile our feelings to the rigour and our reason to the necessity of some plants being made the instrument of destruction to the insect world. Of British plants we have only a few so constructed, which, having clammy joints and calices, entangle them to death. The sundew (Drosera) destroys in a different manner, yet kills them without torture. But we have one plant in our gardens, a native of North America, than which none can be more cruelly destructive of animal life (*Apocynum androsæmifolium*), which is generally conducive to the death of every fly that settles upon it. Allured by the honey on the nectary of the expanded blossom, the instant the trunk is protruded to feed on it the filaments close, and, catching the poor fly by the extremity of its proboscis, detain the poor prisoner writhing in protracted struggles till released by death—a death apparently occasioned by exhaustion alone; the filaments then relax and the body falls to the ground. The plant will at times be dusky from the numbers of imprisoned wretches. This elastic action of the filaments may be conducive to the fertilising of the seed by scattering the pollen from the anthers, as is the case with the berberry; but we are not sensible that the destruction of the creatures which excite the action is in any way essential to the wants or perfection of the plant, and our ignorance favours the idea of a wanton cruelty in the herb; but how little of the causes and motives of action of created things do we know? and it must be unlimitable arrogance alone that could question the wisdom of the mechanism of Him 'that judgeth rightly'; the operations of a simple plant confound and humble us, and, like the handwriting on the wall, though seen by many, can be explained but by One."[1]

---

[1] Knapp, "Journal of a Naturalist," 1838, p. 78.

CHAPTER VI.

OTHER CARNIVOROUS PLANTS.

It is by no means unusual for phenomena of the kind to which the preceding chapters have been devoted, to manifest themselves in a modified form in other organisms. These collateral, or supplementary instances, although of but little moment of themselves, are valuable when taken in their relationship to more complete and perfect examples. In plants to which we shall have occasion to refer, some of the phenomena already described, in association with insectivorous plants, will be repeated, but less intensified; so that such examples will hold an intermediate position between the sundews and plants in which no carnivorous propensities have been traced. As might be expected, some of these supplementary plants approach so closely that they cannot be separated from veritable carnivorous plants, whilst others recede so much, perhaps, as only to be strongly suspected of such proclivities.

The Butterworts are little bog-loving plants, mostly in hilly or mountainous districts, with pale green leaves spreading out like a rosette, and lying flat on the ground. The flowers rise from the centre of the rosette singly, on erect footstalks five or six inches in length. The common butterwort (*Pinguicula vulgaris*) has about eight oblong, thickish leaves in a rosette. They are not more than an inch and a half in length, with a very short footstalk. When fully expanded they fall outwards and lie with their under-surface closely pressed to the surface of the soil. The margins of the leaves have long been known to curve inwards in a peculiar manner, but the reason for this curving had not until recently been at all satisfactorily explained. That they did so was an acknowledged fact, but why they did so was hardly suspected. The surface of the leaves is clammy, and small objects are apt to adhere to them. It having been observed that insects were often found sticking to the leaves led to an investigation of the habits of the plant, and its structure. Here, again, we are indebted to Mr. Darwin for a solution of the mystery,

Fig. 17.—Butterwort (*Pinguicula Lusitanica*).

and must appeal to his work for the details of his discovery.

Thirty-nine leaves with objects adhering to them were sent from North Wales. Thirty-two of these had caught no less than 142 insects. Subsequently nine plants, bearing 74 leaves, were forwarded, and all of these latter, except three young leaves, had insects adhering to them. Another consignment of plants from Ireland had insects on 70 out of 157 leaves. Most of the insects were diptera, or two-winged flies. Circumstances such as these, presented in such a forcible manner to an investigator like Mr. Darwin, would naturally suggest to his mind the inquiry, "why, and how, are these insects caught?"

An examination of the structure of the leaves showed that the upper surface was thickly covered with glands of two kinds. (1) Large circular glands divided into sixteen cells, and supported upon elongated peduncles, or footstalks; and (2) smaller glands of a similar character on shorter peduncles. All the glands secrete a colourless sticky fluid, so viscid that it may be drawn out in threads for a foot or eighteen inches. The edges of the leaves are devoid of glands. Here, then, is sufficient cause for insects and other objects being found adhering to the leaves, as flies may be seen sticking to the old fashioned fly-papers. So that the plant deserves to be considered a "fly-catcher."

We have said that the margins of the leaves have long been known to curve inwards in a peculiar manner. A number of experiments were therefore instituted in order to determine the cause of this curvature, the result being that the movement was produced by irritation of a particular kind. Touching, pricking, or scratching, produced no effect, neither did drops of water, but the continued pressure of such inanimate objects as fragments of glass, or objects furnishing soluble matter, and infusion of meat, caused the leaves to curl. The shortest period at which a decided curvature was observed was about two hours and a quarter. After bending inwards, and remaining so for a short period, the margins open and expand again, say in about twenty-four hours, but are not stimulated to close again for some time. The curvature only takes place in one direction, that is, longitudinally, at either margin, but the apex does not become inflected. When flies or other small objects are placed near the margin of a leaf they are enclosed, or partly so, by the curvature. The glands thus brought into contact with such objects pour out their secretion. But why is the period of curvature so short? We must here permit Mr. Darwin to give his principal explanation.

"We have seen that when large bits of meat, &c., were placed on a leaf, the margin was not able to embrace them, but as it became incurved pushed them slowly towards the middle of the leaf. Any object, such as a moderately-sized insect, would thus be brought slowly into contact with a far larger number of glands, inducing much more secretion and absorption, than would otherwise have been the case. That this would be highly serviceable to the plant we may infer from the fact that Drosera has acquired highly-developed powers of movement, merely for the sake of bringing all its glands into contact with captured insects. In the case of Pinguicula, as soon as an insect has been pushed for some little distance towards the midrib, immediate re-expansion would be beneficial, as the margins could not capture fresh prey until they were unfolded."[1] On the whole, the movements are not accounted for so satisfactorily and completely as in Drosera and Dionœa.

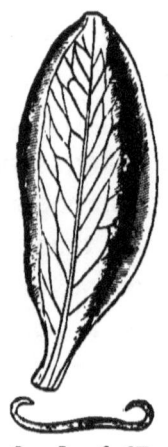

Fig. 18.—Leaf of Butterwort with the edges curved inwards.

In all the British butterworts, and most others, the margins of the leaves naturally curve a little inwards, and this not only serves to prevent insects being washed away by the rain, but also to retain the

---

Darwin, "Insectivorous Plants," p. 379.

secretion. When a number of glands have been excited, the secretion is so profuse that it trickles down towards the incurved edges, but by the curvature is prevented flowing off (fig. 18).

The viscid fluid thrown off by the glands when *not* excited by contact with insects, or other nitrogenous substances, or fluids, is not in the least acid, but when in contact with such substances is invariably acid. This is an important fact to be borne in mind. Flies caused the glands to secrete freely, the secretion being acid. Small bits of roast meat always caused much acid secretion in a few hours. Sugar and starch caused considerable secretion, but it was not acid. Bits of glass excited little or no secretion, that small quantity not being acid. The secretion, when containing animal matter in solution was soon absorbed, causing the glands, which previously were greenish and limpid, to become brownish, and contain aggregated animal matter. No such effect was produced by other substances. The consequence of this acidulation of the secretion, and the interpretation to be attached to the change of colour and aggregation of protoplasm in the glands, have already been explained in the chapter relating to the sundews, and therefore needs not to be repeated here. Suffice it to say that they indicate a process analogous to digestion.

In the course of his experiments Mr. Darwin found

that pollen from flowers, the leaves of other plants, and different small seeds, when they came in contact with the glands caused considerable acid secretion. He considers that albuminous matter would be dissolved out of them and absorbed by the glands. Hence, he says, "we may conclude that the Pinguicula, with its small roots, is not only supported to a large extent by the extraordinary number of insects which it habitually captures, but likewise draws some nourishment from the pollen, leaves, and seeds of other plants which often adhere to its leaves. It is therefore partly a vegetable as well as animal feeder."[1]

Two other species, or reputed species, of Pinguicula which are indigenous to these islands were also examined with similar results. As for *Pinguicula Lusitanica*, the chief difference appeared to be that in this plant the margins when excited were more strongly inflexed, and the inflection lasted for a longer period of time. The glands also apparently were more stimulated to increased secretion when in contact with bodies not yielding soluble nitrogenous matter, than was the case with the common butterwort. In other respects they all agreed, and all alike merit the appellation of insectivorous plants.

Apropos of this we may cite some remarks by one who did not believe or accept the carnivorous theory,

---

[1] Darwin, "Insectivorous Plants," p. 390.

the late Andrew Murray, who has given evidence so much the more valuable from his scepticism on the habits of Pinguicula. "I have been staying," he writes, "in Kinross-shire, where I had an abundance of material to observe, and a fair proportion of both dry and wet weather, so as to see the behaviour of the plant under both conditions. The first thing of which I convinced myself was that, whether it was carnivorous or not, Pinguicula was rightly regarded by Mr. Darwin as coming under the same category as Dionœa and Drosera : it was a fly-catcher and a fly-dissolver—whether it was a fly-digester is a different thing ; but neither on that point any more than on the other can it be separated from them. If the one digests the other will no doubt do so likewise. If the one does not neither will the other."[1] Thus far, then, there is the decided opinion of a very careful observer, that the phenomena exhibited by the Pinguicula are in entire harmony with those of the fly-traps, and that the conclusions arrived at with regard to the latter will also be applicable to the former. "When the insect alights, or is blown on to the leaf," he says, "it gets entangled in the sticky secretion, and it is killed, and speedily killed (long before the curving of the margin of the leaf could

---

[1] Murray in "Gardener's Chronicle," September 19, 1874, p. 354.

have any effect upon it), by the secretion adhering to and closing up the spiracles by which the insect breathes, just on the same principle that a drenching with oil is used in our hospitals, &c., to kill the vermin with which dirty patients or inmates may be swarming on their admission." He proceeds to argue that there is no irritability capable of being exerted by the plant to the injury of the insect, but this injury is confined to the secretion.

Although one could hardly doubt the power of the butterworts not only to catch, but also to digest and assimilate insects, such attributes cannot be claimed for the next group of plants to be noticed. It may be perfectly true that by means of a most elaborate contrivance they are enabled to capture living prey, yet there is no evidence forthcoming that they are able to digest the animals after having caught them. Wherefore, then, it may be asked, an elaborate trap, admirably adapted for the capture of minute organisms, such as are commonly found enclosed within them, if it is all of no use, and when the dainty morsels are captured they cannot be eaten? This question may partly be answered by a brief summary of the results of such investigations as have already been made by Professor Cohn,[1] and Mr. Darwin,[2] and partly reserved as still requiring elucidation.

---

[1] Cohn, "Beitrage zur Biologie der Pflanzen," 1875.
[2] Darwin, "Insectivorous Plants," p. 396, etc.

The bladderworts contain several species, found in different parts of the world, but the principal exami-

Fig. 19.—Bladderwort (*Utricularia vulgaris*).

nations have been made on two British species·

(*Utricularia neglecta* and *Utricularia vulgaris*), fig. 19. These are aquatic, commonly found in dirty ditches, or, as has been said, " remarkably foul ditches." They float freely, not being attached by roots at any period of their existence. The leaves are deeply divided into narrow filaments, each terminating in a short straight point, like a bristle. Small inflated vesicles, or bladders, seated on the leaves, originated the name of bladderwort. It is sometimes stated that these bladders are filled with air, and serve to buoy up the plant in the water. That such an assertion is erroneous may be inferred from the fact that although they often enclose a little bubble of air, they are usually filled with water; that branches float equally well without them ; and that their elaborate construction indicates a much more complex function, whatever that function may be.

Fig. 20.—Bladder of *Utricularia vulgaris*, enlarged.

The chief point of interest is offered by these bladders, the minute structure of which is exhaustively treated in Mr. Darwin's work, but our purpose may be served by a meagre outline. When full grown the bladders are nearly one-tenth of an inch in length, nearly egg-shaped (fig. 20) with the smaller end upwards, and attached obliquely towards the base. The upper, or smaller, end is furnished with six or seven projecting bristles, not unlike antennæ, and, in fact, the bladder resembles some fixed aquatic insect, or some such crustacean as a water-flea anchored upon the leaf. The entrance to the bladder is at the apex, which is closed by a valve opening inwards. The surface of the valve is furnished with numerous glands, and four oblique bristles. The whole inner surface of the bladder is covered with projections, somewhat like stellate hairs, with four arms, two longer directed obliquely backwards, and two shorter ones directed forwards. Each arm encloses generally a minute brown particle in constant motion.

That the use of these bladders is to capture insects may be inferred from the fact of their constant presence when the conditions are favourable. In seventeen bladders (of *Utricularia neglecta*) containing prey of some kind, eight of them contained entomostraca, those lively little crustaceans so common in stagnant water, three enclosed insects, and six the remains of decayed animals past identification. In five bladders, which appeared full, from four to ten crustaceans were found in each. Professor Cohn placed a plant of Utricularia in a vessel of water swarming with crustaceans, and in the morning some of these were observed entrapped in the bladders, in which prison they remained alive for several days.

The entrance to the bladders is effected by bending the valve inwards, which from its elasticity instantly rebounds and cuts off all chance of escape. Occasionally creatures are found fixed in the opening, held by the pressure of the valve, and unable to extricate themselves. No evidence could be obtained that the valves are at all sensitive, as they would not respond to pricking, scratching, or brushing. All that can be affirmed is, that aquatic insects, and like animals enter the bladders by forcing down the valve, and then passing through the slit, which closes after them, and prevents any return. Observations on this process were made by Mrs. Trent, of New Jersey, as witnessed in an American species of Utricularia (*Utricularia clandestina*). She says : " The entrance into the bladder has the appearance of a tunnel net, always open at the large end but closed at the other extremity. The little animals seemed to be attracted into this inviting retreat. They would sometimes dally about the open entrance for a short time, but would sooner or later venture in, and easily open or push apart the closed entrance at the other extremity. As soon as the animal was fairly in, the forced entrance closed, making it a secure prisoner. I was very much amused in watching a water-bear (Tardigrada) entrapped. It went slowly walking round the bladder, as if reconnoitring, very much like its larger

---

[1] Reprinted from "New York Tribune," in "Gardener's Chronicle," 1875, p. 303.

namesake; finally, it ventured in at the entrance, and easily opened the inner door, and walked in. The bladder was transparent and quite empty, so that I could see the movements of the little animal very distinctly, and it seemed to look around as if surprised to find itself in so elegant a chamber; but it was soon quiet, and on the morning following it was entirely motionless, with its little feet and claws standing out as if stiff and rigid. The wicked plant had killed it very much quicker than it kills the snake-like larva. Entomostraca, too, were often captured, Daphnia, Cyclops, and Cypris. These little animals are just visible to the naked eye, but under the microscope are beautiful and interesting objects. The lively little Cypris is encased in a bivalve shell, which it opens at pleasure, and thrusts out its feet, and two pairs of antennæ, with tufts of feather-like filaments. This little animal was quite wary, but, nevertheless, was often caught. Coming to the entrance of a bladder it would sometimes pause a moment and then dash away; at other times it would come close up, and even venture part of the way into the entrance, and back out as if afraid. Another, more heedless, would open the door and walk in, but it was no sooner in than it manifested alarm, drew in its feet and antennæ, and closed its shell. But after its death the shell unclosed again, displaying its feet and antennæ. I never saw even the smallest animalcule escape after it was once fairly inside the bladder.

"So these points were settled to my satisfaction—that the animals were entrapped, and killed, and slowly macerated. But how was I to know that these animals were made subservient to the plant? If I could only prove that the contents of the bladders were carried directly into the circulation, my point was gained. This now was my sole work for several days to examine closely the contents of the bladders. I found the fluid contents to vary considerably, from a dark muddy to a very light transparent colour. Hundreds of these bladders, one after another, were put to the test under the microscope, and I found that, to a greater or less extent, I could trace the same colour that I had found in the bladder into the stem on which the bladder grew, though the observation was not so clear and satisfactory as I could wish. After more critical examination I arrived at the conclusion that the cells themselves, and not their contents, change to a red colour; the stems also take on this colour, so as to make it appear as if a red fluid was carried from the bladders into the main stem, which is not specifically the fact, as far as the observations yet made determine, though the main point, that the contents of the bladders are carried into the circulation, does not seem open to question.

"The next step was to see how many of the

bladders contained animals, and I found almost every one that was well developed contained one or more, or their remains, in various stages of digestion. The snake-like larva was the largest, and most constant animal found. On some of the stems that I examined fully nine out of every ten of the bladders contained this larva or its remains. When first caught it was fierce, thrusting out its horns and feet, and drawing them back, but otherwise it seemed partly paralysed, moving its body but very little; even small larvæ of this species, that had plenty of room to swim about, were very soon quiet, although they showed signs of life from 24 to 36 hours after they were imprisoned. In about 12 hours, as nearly as I could make out, they lost the power of drawing their feet back, and could only move the brush-like appendages. There was some variation with different bladders as to the time when maceration or digestion began to take place, but usually, on a growing spray, in less than two days after a large larva was captured the fluid contents of the bladders began to assume a cloudy or muddy appearance, and often became so dense that the outline of the animal was lost to view.

"Nothing yet in the history of carnivorous plants come so near to the animal as this. I was forced to the conclusion that these little bladders are in truth like so many stomachs digesting and assimilating animal food.

"I have frequently trapped the snake larvæ and seen them enter the bladders. They seem to be wholly vegetable feeders, and specially to have a liking for the long hairs at the entrance to the bladders. When a larva is feeding near the entrance it is pretty certain to run its head into the net, whence there is no retreat. A large larva is sometimes three or four hours in being swallowed, the process bringing to mind what I have witnessed when a small snake makes a large frog its victim."

The trap-like structure of the pitchers, with entrance inwards and no exit; the constant presence of insects in the interior, and quadrifid processes lining the walls, indicating some special function to be performed, would naturally, after the experience of the mode of action in other plants, such as sundews and pitcher-plants, lead to the conclusion not only that insects are caught but also that they are digested. There is, however, no proof that this is the case. The probability even is against it, because there are no secreting glands in the interior of the bladders, and without the outpouring of an acidulated juice we have no experience of a process of digestion taking place. On the other hand, fragments of meat, white of egg, and other substances were inserted in healthy and vigorous bladders, and after two or three days the bladders were cut open and the substances found just in their original condition, without the slightest trace of any digestive process having commenced.

Although there is no proof of digestion there is evidence furnished of absorption by the four-pronged processes. In bladders in which the animal contents were broken up and decayed, the processes exhibited the phenomenon of aggregation, and this we have been led by preceding experience to regard as an evidence of the absorption of soluble animal matter. The same kind of processes in bladders which contained no insects, or in which they were still fresh and unchanged, exhibited no signs of aggregation. As in previous experiments, fluids of a nitrogenous character were applied to the quadrifid processes, and, although without previous sign of aggregation, yet aggregation subsequently took place. The conclusion to be derived from these experiments is, that although the bladders do not digest animal food, yet, after such substances have decayed and become soluble, they are absorbed by the four-pronged processes, causing in them the characteristic aggregation.

There are many bladderworts besides the British species. The Rev. Charles Kingsley recognised them amongst the vegetation of the West Indies. " Our English bladderworts, as everybody knows, float in stagnant water on tangles of hair-like leaves, something like those of the water ranunculus, but furnished with innumerable tiny bladders ; and this raft supports the little scape of yellow snapdragon-like flowers. There are in Trinidad and other parts of South America bladderworts of this type, but those which we found to-day growing out of the damp clay were more like in habit to a delicate stalk of flax, or even a bent of grass, upright, leafless, or all but leafless, with heads of small blue or yellow flowers, and carrying, in one species, a few very minute bladders about the roots; in another, none at all. A strange variation from the normal type of the family, yet not so strange after all as that of another variety in the high mountain-woods, which, finding neither ponds to float in nor swamp to root in, has taken to lodging as a parasite among the wet moss on tree-trunks; not so strange, either, as that of yet another, which floats, but in the most unexpected spots—namely, in the water which lodges between the leaf-sheaths of the wild pines perched on the tree-boughs, a parasite on parasites, and sends out long runners as it grows along the bough in search of the next wild pine and its tiny reservoirs."[1]

Similar curious species of Utricularia were also found by Dr. Gardner in Brazil. One of these especially deserves notice.[2] "Like most of its congeners it is aquatic; but what is most curious is, that it is

---

[1] Kingsley's "At Last," p. 314.
[2] *Utricularia nelumbifolia.*

only to be found growing in the water which collects in the bottom of the leaves of a large Tillandsia that inhabits abundantly an arid, rocky part of the mountain at an elevation of about 5,000 feet above the level of the sea. Besides the ordinary method by seed, it propagates itself by runners which it throws out from the base of the flower-stem. This runner is always found directing itself towards the nearest Tillandsia, when it inserts its point into the water and gives origin to a new plant, which, in its turn, sends out another shoot ; in this manner I have seen not less than six plants united. The leaves, which are peltate, measure upwards of three inches across, and the flowering stem, which is upwards of two feet long, bears numerous large purple flowers." [1]

From this description may be gathered the fact that the species of bladderworts are at least of two kinds, if grouped in accordance with their habits. One group would consist of those which float freely in water, and are truly aquatic ; the other of those species which, like the Brazilian one, are epiphytal or terrestrial, though loving moist places. A third group might almost be constituted of species which live in the crannies of rocks and bear bladders attached to their root-like underground stems. Yet, whatever their peculiar habit may be, the bladders in

---

[1] Gardner, "Travels in Brazil," p. 402.

such as have been examined, even in the dried state, have been found to contain insects. It matters not, even should the bladders be subterranean, their function in all cases is evidently the same, and clearly *not* to cause the plant to float freely in water when so generally present, even in terrestrial species.

As long ago as the year 1858, when examining the species of Utricularia systematically, Professor Oliver remarked : " I may be allowed to express my conviction that in the investigation of the development and general morphology of the bladderworts there is a wide field for extended observation."[1] This was followed by an enumeration of no less than twenty-seven species of Indian Utricularia. Subsequently, the same accomplished botanist published notes upon a number of South American species, two of which are figured with their bladders.[2] There are, indeed, a great number of bladderworts known belonging to Utricularia and allied genera, widely distributed over the globe, and of these nothing is absolutely known of by far the greater number, either of the structure or contents of the bladders or the special habits of the plants themselves.

Mr. Darwin has examined, under the most un-

---

[1] Oliver in " Linn. Journ. and Proceed.," iii., p. 174.
[2] Oliver in "Journ. of Proceedings of Linn. Soc.," iv., p. 169, plate 1.

favourable conditions, the bladders of a few species which had been dried and preserved in herbaria. In one of these (*Utricularia montana*) he found thirty-two bladders on one small branch and seventy-three on another, about two inches in length. In some of the bladders of this species he found animal remains. "The first contained a hairy Acarus, so much decayed that nothing was left except its transparent coat; also a yellow chitinous head of some animal with an internal fork, to which the œsophagus was suspended; also the double hook of the tarsus of some animal; also an elongated, greatly decayed animal; and, lastly, a curious flask-shaped organism having the walls formed of rounded cells" (perhaps the shell of a Rhizopod).[1]

In the Brazilian species, above alluded to by Gardner, he found within one bladder the remains of the abdomen of some larva or crustacean of large size. In a Malayan species,[2] in one bladder there was a minute aquatic larva; and, in another, the remains of some articulate animal. In the bladders of an Indian species[3] were the remnants of Entomostraca; and the bladders of another Indian species[4] contained similar remains. In like manner, in other species from different parts of the world, the bladders enclosed

---

[1] Darwin on "Insectivorous Plants," p. 436.
[2] *Utricularia Griffithii*.   [3] *Utricularia cærulea*.
[4] *Utricularia orbiculata*.

the remains of minute animals. It is a fair inference, therefore, that these bladders, wherever found, are traps to catch unwary insects and minute animals; and as an examination of internal structure shows the presence of similar curious processes to those of the British species having the power of absorption, it may be concluded that decaying animal matter in the bladders is absorbed, and is one natural source of sustenance to the plants.

Very recently Dr. Maxwell Masters has made known the result of some investigations on the Christmas Rose (*Helleborus niger*), a common garden flower, in January, in which he thinks he has discovered an indication, if not of persistent fly-catching, at least of a capacity to assimilate animal food. His remarks are certainly of interest in connexion with the present subject.

"The true petals, formerly called nectaries, of the Hellebore are those peculiar green horns or tubes met with in one or two rows surrounding the stamens, and which secrete a honeyed juice. We suppose that the main object of this secretion is to serve as an attraction to insects to visit the flowers, and so transfer the pollen from one flower to another. We infer this from the fact that the maturity of the anthers and of the stigma is not simultaneous in the same flower, and hence transfer of the pollen to another flower, whose stigma is mature, is a necessity.

If this be correct, it would, of course, be of no advantage to the plant to immolate its insect visitor, as what it would gain in one way it would lose in another. Indeed, we have not found any dead bodies of insects in the tubes of the Hellebore, such as one may find so frequently in the pitchers of Nepenthes or Sarracenia. Hence, then, as a rule it would be of no advantage to the plant to indulge in animal food. But it does not necessarily follow that the plant in question has therefore no such power, or that it does not exert it on occasion.

In the case of the Hellebore, the tubular petals were filled with very finely-chopped cooked meat, leaving some of the tubes unfilled for contrast-sake. The microscopic appearance of the normal petals was noticed, and the reaction of the juice with litmus paper tried.

In the normal tube there are certain cells filled with yellow juice, which is diffused throughout the whole interior of the cell. But after the insertion of the meat, and its retention for some days, the yellow-cell contents were found to be compacted together into a globose ball; at least, in the case of those cells nearest to the meat, those at a distance showed the contents diffused. Moreover, the fibre of the meat may be seen reduced to a pulp, and under the microscope its constituent fibres may be seen disintegrated, and the peculiar striations cha-

L

racteristic of striped muscular fibre even more conspicuous than ordinary. Granular matter and oil globules exist in abundance, and give evidence of partial solution. Some of the same meat kept moist under a bell-glass, side by side with the flowers that formed the subject of experiment, showed little or no trace of disorganisation or putrefaction.

The action of litmus paper is rather puzzling; at first, the juice of the tubes was neutral, or only faintly acid, but after the meat had been allowed to remain for some days, an alkaline reaction was evidenced by the appearance of a blue tinge on previously reddened litmus. For the present, then, we state merely that the muscular fibre was partially dissolved, and that certain changes in the appearance of the cell-contents took place. More than this it would be rash to affirm.[1]

Here terminates our observations on insectivorous plants. It will have been remarked during our progress through the preceding chapters, that there are grades of perfection in the modes by which animal food is obtained. In the more highly developed forms, as in the Sundews, the Venus's Fly-trap, the Butterworts, and the Pitcher-plants, the structure of certain parts is adapted for the capture of insects, which are afterwards covered with an acid secretion

---

[1] "Gardener's Chronicle" (Dr. Masters), April, 1876, p. 470.

and digested in a manner analogous to digestion in animals, the products being absorbed for the benefit of the plant. In the second class there are facilities for the capture of insects in the pitchers of the Side-saddle flowers, and the bladders of the Bladderworts, but the power of digestion is not present, or has not been demonstrated, although there may be the power of absorbing from the decay of imprisoned animals their nitrogenous products. In the first class there are also, in some cases, spontaneous movements designed to aid in the capture of insects which are absent in the second. In the Sundews, the captive, rendered helpless by a viscid secretion, is slowly embraced by sensitive tentacles. In the Fly-trap it is caught by a quick movement, as in a trap, and squeezed to death. In the Butterworts, the capture is accomplished by the profuse tenacious slime, little, or not at all, aided by the curvature of the margins of the leaves; and in the Pitcher-plants there is no spontaneous movement, but an adaptation of parts so as to lure insects to their own destruction.

In order that there may be no doubt that the capture of insects is not "accidental," but a vital process resulting in benefit to the plant, investigations by the aid of the microscope have demonstrated that the absorption of animal matter, or fluids of a like composition, act in a peculiar manner by causing aggregation of protoplasm in the adjacent cells.

Further, that this aggregation has not been observed under any other conditions, and yet has been discovered in all our carnivorous plants, down to the lowest and most uncertain of all, the "Christmas Rose."

In the face of these facts, however reluctant we may at first find ourselves in accepting a theory so contrary to our previous experiences, we must even be "convinced against our will" that there are plants, such as we have described, which are truly carnivorous.

> What's this I hear
> About the new carnivora?
> Can little plants
> Eat bugs and ants
> And gnats and flies?

And, if so, we feel disposed to consider it

> A sort of retrograding;
> Surely the fare
> Of flowers is air,
> Or sunshine sweet;
> They shouldn't eat,
> Or do aught so degrading.[1]

Whatever our feelings may be at first, these must give way under reflection that many yet stranger things than these may yet be revealed. Our knowledge of the operations is as yet very elementary.

---

[1] "Scribner's Magazine," for April, 1875.

Twenty-five years ago Mr. P. H. Gosse wrote thus, in reference to this very subject:—"The curious fact stated by Dr. Lindley, that specimens of the fly-trap fed with atoms of chopped meat have evidently thriven under the stimulating diet, and become more vigorous than others left to the resources of the soil and air, tends to confirm this supposition (of carnivorous habits), however strange it may appear. The frequent occurrence of startling facts, facts at variance with pre-established theories, forbids the philosophic naturalist to speak of any statement, professing to rest on observation, as impossible, merely because it has not been hitherto recognised, or cannot be reconciled with existing knowledge. It was the remark of a sage 'The more I learn, the more do I become convinced that I know nothing.'"[1]

NOTE.—In Gerarde's "Herbal" (p. 601) it is written of the limewort, or viscaria : "The whole plant, as well leaves and stalks, as also the floures are here and there covered over with a most thicke and clammy matter, like unto bird-lime, which if you take in your hands, the sliminesse is such that your fingers will sticke and cleave together, as if your hand touched bird-lime : and furthermore, if flies doe light upon the same, they will be so entangled with the liminesse, that they cannot flie away; insomuch that in some hot day or other you shall see many flies caught by that means. Whereupon I have called it Catch Flie or Limewort."

[1] "Tenby," by P. H. Gosse, p. 209.

CHAPTER VII.

GYRATION OF PLANTS.

IT is only very recently that any regular and systematic investigation has been pursued to ascertain the causes and mode of action in certain phenomena of motion in plants. By whatever names these motions may have been designated, they appear to resolve themselves into modifications of one simple type of movement to which the name of "circumnutation" has been applied. This movement in its ordinary form consists in the revolution of the growing point, which has been described in the following terms :—" If we observe a circumnutating (or revolving) stem, which happens at the time to be bent, we will say towards the north, it will be found gradually to bend more and more easterly until it faces the east, and so onwards to the south, then to the west and back again to the north. If the movement had been quite regular the apex would have described a circle, or rather, as the stem is always growing upwards, a circular spiral. But it generally describes elliptical or oval figures ; for the apex after pointing in any one direction commonly moves back to the opposite side, not, however, returning along the same line."[1] That plants did exhibit motion during growth had been observed, but it was left to more recent times to demonstrate how general this kind of growing movement was, and the direction it assumes. Heliotropism, geotropism, sleep of plants, and circumnutation are expressions for forms of the same phenomena of revolving motion in growing plants.

In the volume which Mr. Darwin has written on the movements of plants, he has distinctly expressed his opinion that apparently every growing part of every plant is continually rotating, though often on a small scale. This movement may be traced in seedlings before they have broken through the ground, and in the extremities of their young roots; in the stems of some climbing plants, and in the tendrils of others; in leaves and leaflets, and in all the growing points.

When seeds commence to germinate they thrust out a small radicle, or rudiment of a root, which at once bends downwards and endeavours to enter the ground. The young rootlet is clad with delicate hairs, which, by the softening of the outer surface and its subsequent hardening helps to attach the plant to the soil. As soon as it emerges from the coat of the seed the young rootlet begins to rotate, and thus

---

[1] Darwin, "On the Movements of Plants," p. 1.

early in its career active spontaneous motion commences. It is true that the oscillations may seem small to the unaided eye, being only about one-twentieth of an inch, or less, in the garden bean, on each side of a central line, but nevertheless distinct and decided. Rootlets were induced to grow down sloping plates of smoked glass, on which their own vibrations were traced, and thus they recorded the extent of their own lateral motion. This rotation of the young radicle seems to be almost universal, whenever not prevented, or until prevented, by the close pressure of the soil. In a loose soil, or when entering worm-holes, this oscillation would doubtless be useful in enabling the root to enter the ground. The bean having been selected for experiment it was shown that the following is an explanation of how the root enters the earth.

"The apex (of the radicle) is pointed, and is protected by the root-cap; the terminal growing part is rigid, and increases in length with a force equal, as far as our observations can be trusted, to the pressure of at least a quarter of a pound, probably with a much greater force when prevented from bending to any side by the surrounding earth. Whilst thus increasing in length it increases in thickness, pushing away the damp earth on all sides with a force of above eight pounds in one case, of three pounds in another case. It was impossible to decide whether the actual apex exerts relatively to its diameter the same transverse strain as the parts a little higher up; but there seems no reason to doubt that this would be the case. The growing part, therefore, does not act like a nail when hammered into a board, but more like a wedge of wood, which whilst slowly driven into a crevice continually expands at the same time by the absorption of water, and a wedge thus acting will split even a mass of rock." [1]

From the seed upwards rises the short rudimentary stem which supports the cotyledons, or seed-leaves. This short stem (if present) raises the cotyledons above the surface of the soil through which it breaks in the form of an arch. When the cotyledons, or seed-leaves, appear, they are at first vertical, with their faces applied to each other; but they soon separate and exhibit the phenomena of rotation. In all the seedlings of dicotyledonous plants examined, the seed-leaves were in constant movement, chiefly in a vertical direction, and most often once up and down in twenty-four hours. In one plant they moved upwards and downwards thirteen times in the course of sixteen hours, but this is unusual. In different species and in different individuals of the same species, there will be variation and gradation in the oscillations. In one species of wood - sorrel, whilst

---

[1] Darwin, "Movements of Plants," p. 77.

one of the seed-leaves moved upwards the opposite one moved downwards. Although the up and down movement was the most common, this was sometimes accompanied by lateral and zigzag oscillations. In a great many cases the seed-leaves sink a little in the forenoon and rise a little in the afternoon, so that they stand rather more highly inclined during the night than at mid-day, when they would be nearly horizontal. The position of the seed-leaves by night and by day was observed by Mr. Darwin in the seedlings of plants in 153 genera. Of these there were twenty-six in which the cotyledons stood vertically at night, or at least sixty degrees above or below the horizon. There were thirty-eight in which the cotyledons (seed-leaves) which at noon were horizontal were at night more than twenty and less than sixty degrees above the horizon. In the remaining eighty-nine the cotyledons did not change their position at night so much as twenty degrees.

Proceeding with its growth, the plumule, or miniature bud of the future stem and true leaves, rises between the cotyledons, or seed-leaves, and as it grows it nutates or rotates as the radicle and the seed-leaves had done.

Before quitting the seedling and its radicle some mention must be made of a subject which occupies an entire chapter of the work to which we have alluded, and that is the sensitiveness of the tip of the radicle.

The radicle of the bean was selected for the majority of the experiments in this connexion, the results of which appear to prove that "when one side of the apex (of a radicle) is pressed by any object the growing part bends away from the object; and this seems a beautiful adaptation for avoiding obstacles in the soil and for following the lines of least resistance. Many organs when touched bend in one fixed direction, such as the stamens of Berberis, the lobes of Dionœa, &c.; and many organs, such as tendrils, whether modified leaves or flower peduncles, and some few stems, bend towards a touching object; but no case, we believe, is known of an organ bending away from a touching object."[1]

That the radicle of many plants, indeed, of most, if not all, are sensitive to pressure continuously exerted upon them, or to injury, and are capable of bending away from it, was shown by many experiments, such as attaching small objects to the side of the tip, touching it with caustic, or cutting off a slice from it. All these interferences seemed to act in a similar manner, causing the tip to diverge from its direct downward course and turn in the direction opposite to the obstruction or injury. This sensitive

---

[1] Darwin, "Movements of Plants," p. 132.

power appeared to be confined to the tip of the radicle for about one-twentieth of an inch. When the part was irritated by contact or slicing, its influence extended upwards for one-third or half an inch, causing the radicle to bend away from the point of irritation in a symmetrical curvature. This occurred sometimes within six or eight hours, and almost always within twenty-four hours of the commencement of the irritation. The curvature often amounted to a right angle, occasionally the tip bent upwards like a hook with its point to the zenith, or it curved and formed a loop or a spire.

The method by which these observations were made was by soaking the beans in water, for about twenty-four hours, then suspending them so that in germinating the radicle might grow downwards freely, and without obstruction. When the radicles were sufficiently protruded, small objects, such as fragments of card or paper, were attached by gum-water, or other cement, on one side of the tip of the radicle, these were then watched for deflection. In other instances slices were cut with a razor from one side of the tip, so as to cause continued irritation. In all these instances the radicles responded, and turned away in the opposite direction to the source of irritation. In cases where the tips were only temporarily irritated, by striking or pricking, no deflection occurred. Other tips were touched lightly on one side with dry nitrate of silver. The injury caused by the caustic was permanent, and the tips turned away in the opposite direction usually with more certainty than when objects were attached.

For the proper observation of these movements the radicle should be developed at the normal rate. If subjected to high temperature, so that it grows rapidly, or checked so as to germinate slowly, as in winter, the irritability is much less pronounced. The common garden pea, treated in the same manner, was found to be even more sensitive than the bean.

A condition of irritability, or sensibility, had been previously demonstrated in radicles, but this always took place higher up, and not at the tip. In such cases the radicle always turns towards the touching object, and not away from it. This movement is of a kindred nature to that exhibited by tendrils, which turn towards and embrace the touching object. But the sensibility of the *tip* seems to be of quite a distinct character, and one which would prove of considerable service to the plant. As Mr. Darwin says : " The direction which the apex takes at each successive period of the growth of a root ultimately determines its whole course ; it is, therefore, highly important that the apex should pursue from the first the most advantageous direction ; and we can thus understand why sensitiveness to geotropism, to contact, and to moisture, all reside in the tip, and why the tip determines the upper growing part to bend either from or to the exciting cause. A radicle may be compared with a burrowing animal, such as a mole, which wishes to penetrate perpendicularly down into the ground. By continually moving his head from side to side, or circumnutating, he will feel any stone or other obstacle, as well as any difference in the hardness of the soil, and he will turn from that side; if the earth is damper on one than on the other side he will turn thitherward as a better hunting-ground. Nevertheless, after each interruption, guided by the sense of gravity, he will be able to recover his downward course, and to burrow to a greater depth."[1]

From seedlings we are led to mature plants, and here again we encounter systematic rotatory movement so universal in its character that it is doubtful if it does not exist more or less in all plants.

A hybrid raspberry, about a foot high, that was growing vigorously, was watched in the month of March. During the morning the growing point of the stem almost completed a circle, and then deflected to the right. In the afternoon it reversed its course, and continued to move in that direction for forty hours.

Some young ivy-plants were seen to rotate their

---

[1] Darwin, "Movements of Plants," p. 200.

growing points at a slow rate, and over a small space during six successive days.

The stems of some plants only describe one large ellipse during twenty-four hours, others describe several smaller ones. A plant which describes but one ellipse during one day, may on the next describe two smaller ones. Sometimes the course is almost circular, but generally elliptical, and the ellipse may be broad or very narrow. Whatever the figure may be it is not a regular one, but varied by loops or zigzag vibrations in other directions.

In coniferous trees it was observed by Dr. Masters in 1878[1] that the leading shoot has a rotatory movement, and this has subsequently been confirmed by him, in *Abies Nordmanniana*. The same authority has also directed attention to other movements which affect the leaves, and relate to modified forms of heliotropism. "Some of the silver firs," he says, "are endowed with a power of motion by means of which the leaves are raised or lowered. M. Chatin asserts that in the middle of the day the plant (*Abies Nordmanniana*) has a predominant green hue, but when the light is more diffused, as in the evening or early morning, then the plant assumes a milky-white appearance. This appearance is due to the elevation of the glaucous

---

[1] "Gardener's Chronicle," 1878, pp. 247 and 826.

under-surface of the leaves. The movements in question are so obvious during the period of active vegetation that no doubt can exist on the subject. Unlike M. Chatin, however, I have observed that the white hue of *Abies Nordmanniana* is more conspicuous when the branches are exposed to the full rays of the sun. The same remark holds good of other species."[1]

The tips of runners, or stolons, of such plants as the garden strawberry, exhibit just the same rotating movements as stems. The same may be said also of the flowering stems of various plants, such as that of the rape, as observed by Sachs; wood-sorrel, as observed by Darwin and others, to which the last author has alluded. A larger number of observations has been made on the rotation, or circumnutation of leaves; and, as the phenomena are similar in all, illustrations of this latter kind will suffice.

The leaves of the cabbage rise at night and fall by day, an irregular ellipse being formed every twenty-four hours. The leaves of the Swedish turnip draw together so much in the evening, that, according to Mr. Stephen Wilson, " the horizontal breadth diminishes about 30 per cent. of the daylight breadth."

In the common bean (*Vicia faba*) the whole leaf and the terminal leaflets pass through regular well-

---

[1] "Journal of Linnæan Society," vol. xvii., p. 550.

marked diurnal movements, rising in the evening and falling during the latter part of the night, or early morning, whilst during the middle of the day they rotate around the same small space.

More complex movements take place in Venus's Flytrap (*Dionæa*). In a young state the two lobes of the leaf are pressed closely together. In the evening one of these young leaves formed an ellipse in the course of two hours. An older leaf did not rotate plainly. A young and unexpanded leaf was carefully watched with a micrometer. It moved onwards generally by rapid jerks. After each jerk the apex drew itself backwards slowly for part of the distance that it had advanced, and then, slowly afterwards, made another jerk forwards. Four conspicuous jerks forwards, with slower retreats, were on one occasion seen to occur in exactly one minute. Sometimes the apex remained quite motionless for a short period. An older leaf was tested in a similar manner at a lower temperature, the apex oscillated forwards and backwards in the same way, but the jerks forward were less, and the motionless periods longer. During one of these motionless periods a wax taper was held close to the leaf. After ten minutes violent oscillations commenced, perhaps owing to the stimulus of the warmth of the taper. The light was then removed, and in a short time the oscillations ceased. Looked at again after a period of an hour and a half, and it was again oscillating. The plant was taken back into the hothouse, and in the following morning was seen to be still oscillating, but not vigorously.

A cyclamen was also watched under the usual conditions for three days. On the first day the leaf fell more than afterwards. On all three days it fell from early morning until about seven in the evening, and from that time it rose during the night, with a slightly zigzag movement. Although the whole distance travelled was considerable, yet the motion would hardly attract attention, or be observed, unless some method of tracing or measuring the movements were adopted.

Plants of seakale (*Crambe maritima*) were also under observation. In the first instance, a leaf nine inches long was selected. Its apex was in constant movement, but this could hardly be traced, from being so small in extent. A more vigorous young plant with four leaves was then selected. One of the leaves was specially watched and found to be continually rotating, and its movements were distinctly traced. One of the leaves changed its course at least six times in fourteen hours.

The leaves of a camellia were also observed. These leaves are firm and leathery, with short footstalks, so that but little apparent rotation was expected. Nevertheless the apex of a leaf changed its course completely seven times in eleven hours, but moved to only a very small distance. On both days the leaf rotated in the forenoon, fell in the afternoon, and then rose, falling again during the night or early morning.

Peculiar movement in the frond of a fern has been recorded by Professor Asa Gray. "A tuft of *Asplenium trichomanes*, gathered last autumn in the mountains of Virginia, is growing in the house of Mr. Loomis, in a glass dish. About two months ago he noticed that one of the fronds,—a rather short and erect one, which is now showing fructification,—made quick movements alternately back and forth, at right angles to the frond, through from $20°$ to $40°$, whenever the vessel was brought from its shaded situation into sunlight or bright daylight. The movement was more extensive and rapid when the frond was younger. When I first saw it (on 23rd January), its compass was within $15°$, and was about as rapid as that of the leaflets of *Desmodium gyrans*. It was more rapid than the second hand of a watch, but with occasional stops in the course of each half-vibration. This was in full daylight, next a window, but not in sunshine. No movement had been observed in the other fronds, which were all sterile and reclining, with the exception of a single one which was just unfolding, in which Mr. Loomis thinks he has detected incipient motion of some kind."[1] Subsequently to the publication of this notice, Mr. Loomis furnished the following additional particulars. "Four other fronds starting from two different roots exhibit motion, but in less degree than the one first noticed. These are not new fronds, but old ones which were fully developed as to size when taken up, but have fruited since transplanting. It seems to me that the motion is confined not only to the fruitful fronds, but to the period of fructification, since these four fronds have been subjected to the same conditions as the first, but have exhibited motion only since the fruiting began. The stimulus of artificial light is sufficient to excite motion in the fronds for a few minutes, but after the lapse of five or six minutes the motion ceases, and is not resumed. I have noticed that the end of the frond does not describe a straight line, but it moves in a long and very narrow ellipse with the hands of a watch. The motion is more vigorous and through a larger arc in the middle of the day."[2] Interest attaches to this narrative on account of the very few instances of spontaneous movement as yet recorded in the higher cryptogams; and it would be well if it stimulated closer observation of the ferns and their allies.

There is one other point which comes within the limits of this chapter, and may be referred to here

---

[1] "Botanical Gazette," 1880, p. 27.   [2] "Ibid.," p. 43.

with advantage. There are some plants which as the flowers fade point their ovaries downwards, and then by the curvature or lengthening of the peduncle, these ovaries are made to enter the ground and mature their seeds in the earth. The explanation which is offered to account for such movements is that these parts are more than ordinarily sensitive to gravitation, and that it is "geotropism" by which such phenomena should be called, because of their turning to the earth, as "heliotropism" is applied to those which turn towards the sun. Whatever the explanation may be, the phenomena are interesting as exhibiting a curious type of movement of plants.

We will commence with a species of clover (*Trifolium subterraneum*) which is indigenous to the south of England, and therefore of more interest than an exotic would be. For the details of its burying propensities we must again follow the lucid narrative of Mr. Darwin, but somewhat condensed. "The flower-heads of this plant produce only three or four perfect flowers. All the other flowers are abortive and modified into rigid points. After a time five long elastic claws which represent the divisions of the calyx are developed on their summits. As soon as the perfect flowers wither they then bend downwards, as the peduncle stands erect, and closely surround its upper part. The imperfect flowers, which are the central ones in the flower-heads, ultimately follow, one after another, the same course. Whilst the perfect flowers are bending, the whole peduncle (flower-stalk) curves downwards, increasing in length until the flower-head reaches the ground. Nineteen upright flower-heads, arising from branches in all sorts of positions were marked, and after twenty-four hours six of them were vertically depressed, having travelled through 180°. Ten were extended horizontally, and these had passed through about 90°. Three very young peduncles had as yet only moved a little downwards.

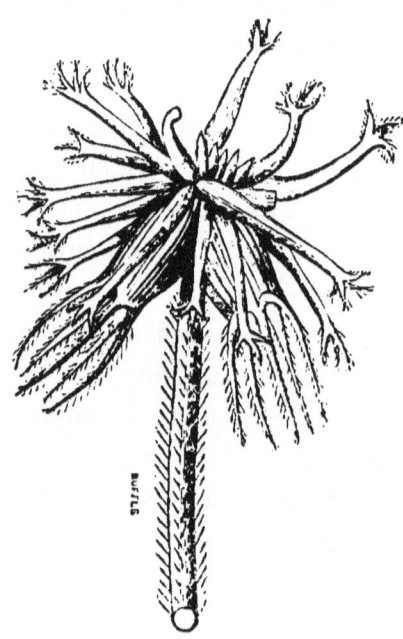

Fig. 21.—Trifolium subterraneum fruit.

"When the flower-heads reach the ground, the younger imperfect flowers in the centre are pressed together in the form of a cone, whereas the perfect, and the outer imperfect flowers are reflexed and closely surround the footstalk. Thus the form they assume is adapted to offer as little resistance as possible in penetrating the

GYRATION OF PLANTS. 167

soil. The flower-heads are able to bury themselves in garden mould or sand. The depth to which they penetrate, measured from the surface to the back of the head, is from a quarter to half an inch. With plants in the house a head partly buried itself in six hours. After three days only the tips of the reflexed calices were visible. In six days the whole had disappeared. Plants growing out of doors are believed to bury their flower-heads in a much shorter time. Only a few of the flower-heads which from their position are not able to reach the ground yield seeds, whereas the buried ones never failed to produce as many seeds as there had been perfect flowers.

"It is unnecessary to enter into all the details of observations on the movement of the footstalk, from the time it begins to bend until the flower-head is buried in the ground. Suffice it to say, that throughout this period oscillation was going on. A peduncle was watched during fifty-one hours, whilst in the act of burying itself in a heap of sand. When buried so that the tips of the sepals alone were visible, it was rotating. When the flower-head had completely disappeared beneath the sand it was still rotating. Any one who will observe this process will be convinced that the rocking movement due to the rotation of the peduncle bears an important part in the act. Considering that the flower-heads are very light, that the peduncles are long, thin, and flexible, and that they arise from flexible branches, it is incredible that an object as blunt as one of these flower-heads could penetrate the ground by means of the growing force of the peduncle, unless it were aided by the rocking movement. After a flower-head has penetrated the ground to a small depth, another and efficient agency comes into play; the central rigid aborted flowers, each terminating in five long claws, curve up towards the peduncle, and in doing so can hardly fail to drag the head down to a greater depth, aided as this action is by the circumnutating movement, which continues after the flower-head has completely buried itself. The aborted flowers thus act something like the hands of a mole, which force the earth backwards and the body forwards." [1]

Another instance, equally remarkable, is that of the "ground-nut," or "ground-pea" (*Arachis hypogæa*) which is cultivated in all tropical countries. After the flowers fall the stalk of the ovary elongates itself considerably, bends downwards, and the ovary is buried in the ground, where the pod is matured. It is said that flowers which grow too high on the plant to reach and bury themselves in the ground, do not produce seeds. The same phenomenon of rotation or oscillation is plainly visible in the ovaries directed towards the ground as in the previous example. Any

---

[1] Darwin, " Movements of Plants," p. 514.

sane person would arrive at the conclusion that this subterranean habit is of some service to the plant; that all this elaborate adaptation for burrowing is designed for some purpose, and is not merely accidental; and that the smaller contributories to the act, whilst they aid in its consummation, do so to that intent.

We have endeavoured to show, in this chapter, that the rotation, or oscillation, of the growing parts of plants is a phenomenon which is exceedingly common; that it exhibits itself first in the young radicle soon after it emerges from the seed; that it is very prevalent in the cotyledons or seed-leaves, and in the plumule or growing point; that the young parts of all plants evidence its presence to a greater or less extent; that leaves possess the faculty as well as cotyledons; and, finally, that not only do flower-stems oscillate, but that after the flowers have withered the same phenomenon accompanies the ovary of those plants which mature their seeds beneath the surface of the soil.

We have, by the way, illustrated the peculiarly sensitive character of the tip of the radicle, and endeavoured to indicate its service to the plant. It has only been our aim to summarise what is known of these phenomena, and to present the most striking features as subjects for thoughtful reflection, and as incentives to closer observation.

CHAPTER VIII.

HELIOTROPES, OR SUNFLOWERS.

THE designation which we have adopted for this chapter is simply intended to intimate that we desire to include under it such observations as we purpose to make on plants which conspicuously turn themselves towards or from the sun, or parallel to its course,—

> As the sunflower turns on its god as he sets
> The same look which it turn'd as he rose,

is somewhat amplified so as to include a greater variety of movements, more or less dependent on light. Thus, for instance, the American Compass-plant, which is affirmed to present the edges of its leaves duly north and south, although not truly heliotropal, could not be omitted. This plant has evidently been long familiar to the hunters of the prairies, on account of the direction of its leaves, although it was not made known to the scientific world until 1842.

Captain Mayne Reid calls it the Polar-plant, for he says of it, "We had a guide to our direction, unerring as the magnetic needle. We were traversing the region of the Polar-plant, the planes of whose leaves at almost every step pointed out our meridian. It grew upon our track and was crushed under the hoofs of our horses as we rode onwards."[1] Under the same name it is referred to by Burton: "Whilst in the damper ground appeared the Polar-plant,—that prairie compass, the plane of whose leaf ever turns towards the magnetic meridian."[2] The *Times* correspondent with the Prince of Wales in Canada alludes to it as the Compass-weed: "Fortunately none go to the prairies for the first time without being shown, in case of such mishaps, the groups of Compass-weed which abound all over the plains, and the broad flat leaves of which point due north and south with an accuracy as unvarying as that of the magnetic needle itself."[3] Lieut. J. W. Albert says: "The prairie was yet what is called rolling, the flat bottoms were covered with the rosin, weed, or Polar-plant (*Silphium laciniatum*), whose pinnate-parted leaves have their lobes extending like fingers on each side of the midrib. It is said that the planes of the leaves of this plant are coincident with the plane of the meridian; but those I have noticed must have been influenced by some

---

[1] "The Scalp-Hunters" (1852), p. 206.
[2] "The City of the Saints," by R. F. Burton (1861), p. 60.
[3] 1861, p. 300.

local attraction that deranged their polarity."[1] Another officer in the United States army calls it the "Pilot-weed." Hence it is evident that the belief is widely prevalent that its leaves affect polarity.

> Look at this delicate plant that lifts its head from the meadow,
> See how its leaves all point to the north, as true as the magnet ;
> It is the compass-flower, that the finger of God has suspended
> Here on its fragile stalk, to direct the traveller's journey
> Over the sea-like, pathless, limitless waste of the desert.[2]

And, in confirmation of this description by the poet, it is stated, on authority, that " repeated observations upon the prairies, with measurements by the compass of the directions assumed by hundreds of leaves, especially of the radicle ones, have shown that as to prevalent position, the popular belief has a certain foundation in the fact." It has also been found that the anatomical structure of the leaves corresponds to this position. Since the leaves tend to assume a position in which the two faces are about equally illuminated by the sun, it has been observed that the stomata are about equally abundant on both sides, and the arrangement of " palisade cells " of both strata are nearly the same. Sir Joseph

---

[1] Appendix to " Notes with the Army of the West" (1848), p. 388.   [2] Longfellow's " Evangeline."

Hooker says, that when traversing the prairies with Professor Asa Gray, in 1877, he watched the position of the leaves of many hundred plants from the window of the railway car, and after some time persuaded himself that the younger, more erect leaves especially, had their faces parallel approximately to the meridian line. At the same time he says that he convinced himself that "the flower-heads of various of the great Helianthoid Compositæ, such as that which we call the 'sunflower,' that grew in hosts on the prairies *did* follow the sun's motion in the heavens to a very appreciable degree, their morning and evening positions being reversed."[1]

It has been truly said that no one can look at the plants growing on a bank, or on the borders of a thick wood, and doubt that the young stems and leaves place themselves so that the leaves may be well illuminated. Whoever has placed half a dozen plants on a window-sill knows well enough how soon all the leaves and extremities of the branches will be directed towards the window. The tendency of plants to turn to the light is known to every cultivator, but this tendency is more strongly developed in some plants than in others, and in some parts of the same plant than in others. The types of heliotropism, in its broadest sense, are four, and to these, for the sake

---

[1] J. D. Hooker, in "Gardener's Chronicle," January 15, 1881.

of precision, distinct terms have been applied. In the first place, there are plants which when exposed to a strong lateral light turn speedily towards it. This is true *heliotropism*, or turning towards the sun. Then, secondly, there are plants which, when exposed in a similar manner to a bright side-light, manifestly and speedily turn from it. This has been called "negative heliotropism," but the term which Mr. Darwin has employed to designate such movements is *apheliotropism*. Thirdly, there are plants which on exposure to light, when sufficiently intense, place their leaves transversely to the direction whence the light proceeds, and this has been called "transverse heliotropism," or, as Mr. Darwin prefers to call it, *diaheliotropism*. And finally, there are those plants which direct their leaves by rising, or falling, or twisting, so that they may be less intensely illuminated, and these movements, which are sometimes called "diurnal sleep," the same author designates as *paraheliotropism*. It is clear that if grouped in accordance with their purposes the first and third of these types are allied, as also are the second and fourth. In "heliotropism" and "transverse heliotropism" the object is to turn into and take advantage of the light. In "negative heliotropism" and "diurnal sleep," on the contrary, the object seems to be turning from, or shunning as much as possible, the direct influence of the light. We have given above the four terms by which it is proposed to designate these four movements, but for our purposes we shall make as little use of them as possible.

True heliotropism prevails very extensively amongst the higher plants, but there are some remarkable exceptions; as, for instance, in the "carnivorous plants." The sundews and side-saddle flowers exhibit in their leaves and pitchers no trace of heliotropism. The stems, tendrils, and rootlets of climbing plants are often opposed to heliotropism, or negatively heliotropic, whilst the leaves, on the contrary, have a general tendency to turn towards the sun. Most seedlings are strongly heliotropic, even though afterwards, as they grow up, they become either uninfluenced by the direct light, or, in some cases, turn away from it. Evidently some of the lowest forms of vegetable life seek the light. Strasburger says that the cells of *Hæmatococcus*, a simple unicellular alga, moved to a light which only just sufficed to allow middle-sized type to be read.[1] It is well known how the species of *Oscillaria* congregate towards the light. Some Desmids and Diatoms exhibit the same propensity, the latter especially will rise to the surface and form a scum on the water in the full blaze of sunlight.

In the experiments necessary for demonstration

[1] "Wirkung des Lichtes auf Schwarmporen," p. 52.

of the fact of heliotropism, plants had to be selected which were known to be peculiarly sensitive, and also readily available and of convenient size. The young seedlings of the canary grass (*Phalaris canariensis*) were selected as fulfilling the conditions, just after they appeared above the soil. Of course there was nothing particularly sensational in such experiments; they did not make a great show, and produced no very startling results, nothing greatly beyond the confirmation of what was previously known. Yet they are not without their interest, and especially those which were made with the view of testing the small amount of light necessary to induce heliotropism. A pot of these seedlings, which had been raised in the dark, was placed in a dark room at the distance of four yards from a small lamp. Within three hours the shoots were perhaps slightly curved to the light, but so little as to be doubtful. In rather more than seven hours all were plainly bent towards the light. Now, the light was so feeble at the distance named that the Roman figures could not be distinguished on the white face of a watch, and no shadow was cast by a pencil held upright on a white card, so that the amount of light diffused must have been exceedingly small, and yet it did not fail to exert its power after upwards of seven hours' exposure. Similar experiments were performed at greater and at less distances, and in a variety of

HELIOTROPES, OR SUNFLOWERS. 177

ways, but with substantially similar results. The movement is, of course, very slow where the light is dim, but more rapid where the light is more intense. Placed before a bright lamp the tips of the shoots were all curved at right angles towards it in two hours and a quarter. From other experiments it was determined that seedlings of this grass commenced travelling in the direction of the lamp within from six to ten minutes after their first exposure. Also that the rate of progression was irregular, sometimes almost stationary for ten minutes, and then onwards again. It may be mentioned in this connexion that a series of observations was made in order to determine how long the influence of light would continue to be exerted after the source of light was obscured. It was found that the young shoots would continue to bend in the same direction as that from which the illumination proceeded for from a quarter to half an hour after the light was extinguished.[1]

Movements in the direction of the light have been so universally observed, that heliotropism has been accepted as a fact, independently of any recent observations. The experience of any one who has been concerned with vegetation will furnish numerous instances of such phenomena. The farmer, the nurseryman, the gardener, and even the field labourer,

---

[1] Darwin, "Movements of Plants," pp. 457 to 463.

will have something to tell of the movements of leaves and flowers towards or away from the light. It is unnecessary, therefore, to multiply examples to prove that which is generally accepted. Each, perhaps, will have his own theory of the reason why these phenomena are exhibited, and here, too, the tendency will be in the same direction, namely, that the growing point, or the leaf, or the flower is turned to the light to secure some advantage to the plant. In fact, that heliotropism acts for a purpose, and not as a blind chance; that it is one of the means adopted to secure certain ends, and this appears to be incontrovertible. "It is of more importance to insectivorous plants to place their leaves in the best position for catching insects than to turn their leaves to the light, and they have no such power. If the stems of twining plants were to bend towards the light they would often be drawn away from their supports, and they do not thus bend." And thus, with other exceptions, there is usually an efficient reason which will suggest itself to the thoughtful mind as a good and sufficient cause.

That modified form of heliotropism which has been called "transverse heliotropism," or *diaheliotropism*, has, perhaps, been less observed, because less conspicuous than true heliotropism. The best examples of this form must be sought amongst seedlings, the seed-leaves of which are extended horizontally. "If," says Mr. Darwin, "the two (seed-leaves) are placed in the line of entering light, the one farthest from it rises up and that nearest to it often sinks down; if placed transversely to the light they twist a little laterally, so that in every case they endeavour to place their upper surfaces at right angles to the light. So it notoriously is with the leaves on plants nailed against a wall or grown in front of a window. A moderate amount of light suffices to induce such movements; all that is necessary is, that the light should steadily strike the plants in an oblique direction."[1]

It must be borne in mind that in determining the range of sleep in plants, an artificial limit of rise or depression had to be accepted, and this was taken at 60° above or below the horizon. Consequently there would be a number of plants which would not be ranged with "sleepers," simply because the elevation or depression of their leaves or leaflets uniformly fell short of 60°. It is probable that some of these instances would fall appropriately under "transverse heliotropism" if their movements were accurately determined. There can be no doubt that all these motions have an intimate relationship to each other, and it is contended that all are modifications of one motion, namely, that of circumnutation.

---

[1] Darwin, "Movements of Plants," p. 439.

As heliotropism, or turning towards the sun, is so common a phenomenon, it may be anticipated that negative heliotropism, or turning from the sun, is rare. This latter is so distinctly the case, that Mr. Darwin could only record two instances which had come under his personal observation. The first of these was in the case of an exotic trumpet-flower (*Bignonia capreolata*). It was the tendrils of this climbing plant which exhibited a dislike of the light. This plant was placed in a north-east window, protected on all other sides from the light. It had a pair of tendrils which stood almost vertically upwards. In fifty minutes both tendrils had felt the full influence of the light, for they moved straight away from it for nearly three hours, when they rotated a little and then continued to move away. By a late hour in the evening they had moved so far that they were turned in a direct line from the light. During the night they returned a little in the opposite direction, but, on the following morning, they again moved away from the light, interlocking with each other, but still pointing from the light.

The other instance was that of the cyclamen, and the record is so circumstantial and interesting that we are tempted to give it entire. "Whilst this plant is in flower the peduncles stand upright, but their uppermost part is hooked, so that the flower hangs downwards. As soon as the pods begin to swell the peduncles increase much in length, and slowly curve downwards, but the short, upper, hooked part, straightens itself. Ultimately, the pods reach the ground, and if this is covered with moss or dead leaves, they bury themselves. We have often seen saucer-like depressions formed by the pods in damp sand or sawdust, and one pod (three-tenths of an inch in length) buried itself in sawdust for three-quarters of its length. The peduncles can change the direction of their curvature, for if a pot with plants having their peduncles already bowed downwards, be placed horizontally, they slowly bend at right angles to their former direction towards the centre of the earth. We, therefore, at first, attributed the movement to geotropism (turning towards the earth), but a pot which had lain horizontally with the pods all pointing to the ground, was reversed, being still kept horizontal, so that the pods now pointed directly upwards ; it was then placed in a dark cupboard, but the pods still pointed upwards after four days and nights. The pot, in the same position, was next brought back to the light, and after two days there was some bending downwards of the peduncles, and on the fourth day two of them pointed to the centre of the earth, as did the others after an additional day or two. Another plant, in a pod which had always stood upright, was left in the dark cupboard for six days ; it bore three peduncles, and only one became within this time at all bowed downwards, and that doubtfully. The weight, therefore, of the pods is not the cause of the bending down. This pot was then brought back into the light, and after three days the peduncles were considerably bowed downwards."[1] The further description only relates to the oscillating or rotating movement by means of which the pods excavated for themselves saucer-like depressions in sand or sawdust, or were enabled to bury themselves amongst moss, &c.

The phenomenon of "diurnal sleep" may be illustrated by two familiar examples. The attributes of this movement consist in turning sideways to the light, so that the edges of the leaves or leaflets being directed towards the sun the surfaces should escape its direct influences. The false acacia, or thorny acacia (*Robinia pseudacacia*), is a common tree in this country, not uncommon in towns, but more in favour on the Continent than with us. The leaves are compound; that is, they consist of a row of leaflets arranged along each side of a common footstalk. It is well enough known that these leaflets at night are pointed vertically downwards, whereas under a moderate light they are almost horizontal. When the sun shines brightly on these leaves another

---

[1] Darwin, "Movements of Plants," p. 433.

movement takes place, for the leaflets rise upwards and present their edges to the light. This is the movement called "paraheliotropism," and is made in the reverse direction to the nocturnal motions of the same organs.

The other example is also that of a plant which has been alluded to as going to sleep at night, viz., the wood-sorrel (*Oxalis acetosella*). In this plant the leaves are depressed under the influence of a strong light, just in a similar manner to their nocturnal movements when deprived of light. The motion in one case is not distinguishable from that in the other. It is stated on the authority of Professor Batelin that the leaflets of this plant may be exposed to the sunlight daily for many weeks, and not suffer from the exposure if they are allowed to depress themselves. But if the depression of the leaflets is prevented, by any mechanical means, they lose their colour, and wither in two or three days. Yet the duration of the leaves naturally is about two months, when subjected only to diffused light. Under such latter conditions they do not become depressed during the day.

That the leaves of a tree, or a plant, should inherit the faculty of self-preservation by this simple expedient, is one of the "curiosities of vegetation." That it may be something more, we leave to the conviction which it may present to thinking minds.

CHAPTER IX.

TWINERS AND CLIMBERS.

WHOEVER has read the records of travellers in tropical forests will have been struck with the constant recurrence of some reference to climbing plants, even should they fail to remark the general allusion to climbers and twiners as important features in tropical vegetation. We take up the first book at hand, and in it we read, as follows :—" Below, the tree trunks were everywhere linked together by sipos; the woody, flexible stems of climbing and creeping trees, whose foliage is far away above, mingled with that of the taller independent trees. Some were twisted in strands like cables; others had thick stems contorted in every variety of shape, entwining snake-like round the tree-trunks, or forming gigantic loops and coils among the larger branches; others again, were of zigzag shape, or indented, like the steps of a staircase, sweeping from the ground to a giddy height. It interested me much afterwards to find that these climbing trees do not form any particular family. There is no distinct group of plants whose especial habit is to climb, but species of many, and the most diverse families the bulk of whose members are not climbers, seem to have been driven by circumstances to adopt this habit. There is even a climbing genus of palms (*Desmoncus*), the species of which are called in the Tupi language, 'Jacitara.' These have slender thickly-spined and flexuous stems, which twine about the taller trees from one to another, and grow to an incredible length. The leaves, which have the ordinary pennate shape characteristic of the family, are emitted from the stems at long intervals, instead of being collected into a dense crown, and have at their tips a number of long recurved spines. These structures are excellent contrivances to enable the trees to secure themselves by in climbing, but they are a great nuisance to the traveller, for they sometimes hang over the pathway, and catch the hat or clothes, dragging off the one or tearing the other as he passes. The number and variety of climbing trees in the Amazon forests are interesting, taken in connexion with the fact of the very general tendency of the animals, also, to become climbers."[1]

In a similar manner the Rev. Charles Kingsley was impressed with these plants in the forests of the West Indies. "Around your knees are probably

---

[1] "The Naturalist on the Amazons," by H. W. Bates, p. 17.

Mamures, with creeping stems and fan-shaped leaves, something like those of a young coco-nut palm. You try to brush through them, and are caught up instantly by a string or wire belonging to some other plant. You look up and around, and then you find that the air is full of wires—that you are hung up in a network of fine branches belonging to half a dozen different sorts of young trees, and intertwined with as many different species of slender creepers. You thought at your first glance among the tree stems that you were looking through open air; you find that you are looking through a labyrinth of wire rigging, and must use the cutlass right and left at every five steps."[1]

To these we will add only one other description of the characteristics of a virgin forest. "Its striking characteristics were, the great number and variety of the forest trees, their trunks rising frequently for sixty or eighty feet without a branch, and perfectly straight; the huge creepers, which climb about them, sometimes stretching obliquely from their summits like the stays of a mast, sometimes winding around their trunks like immense serpents waiting for their prey. Here, two or three together, twisting spirally round each other, form a complete living cable, as if to bind securely these monarchs of the forest; there,

---

[1] "At Last," by C. Kingsley, p. 157.

they form tangled festoons, and, covered themselves with smaller creepers and parasitic plants, hide the parent stem from sight."[1]

In this temperate clime of ours we know nothing of the gigantic climbers of the tropical forests, but we have many plants with a similar habit, on a small scale, quite sufficient to give us an interest in the phenomena concerned in the twining process. It was in 1865 that Mr. Darwin's memoir of the habits of climbing plants first made its appearance,[2] and, as since revised, is now the text-book on the subject.

In this work scandent plants are divided into four classes, applicable alike to our purpose:—(1) *Twiners*, those which twine spirally round a support unaided by any other movement; (2) *Climbers*, endowed with irritable organs, which, when they touch any object, clasp it; (3) *Scramblers*, which ascend merely by the aid of hooks; and (4) *Root-climbers*, which ascend by means of rootlets attached to their support. Of these four classes the first and second are of most importance, being by far the most numerous, and true climbers; the third and fourth being pseudo-climbers.

We could not desire a more familiar or better illustration of a "twiner" than the hop (*Humulus*

---

[1] "Travels on the Amazon," by A. R. Wallace, p. 23.
[2] "Journal of the Linnæan Society," vol. ix.. p. 1.

lupulus), and by a little attention to the habits of this plant we may comprehend the general principle on which all twining plants ascend. The first two or three joints or internodes of the hop, after it emerges from the ground, are straight and stationary. We use the term "internode" as expressing that portion of a stem which lies between one node or knot and the next; that is, between the point where one leaf, or pair of leaves, spring from the stem, and the like point next above it. After the two or three straight joints, or internodes, of the young shoot of the hop, another one grows, which, whilst still young, is seen not to be motionless, like its predecessors, but to bend on one side and move slowly round, or rotate, like the hands of a watch, with the sun. We have already become familiar with this kind of rotation in plants, but must be prepared to meet with it in twining plants in a more exaggerated degree. The average rate of rotation during the day in hot weather was found by experiment to be two hours and eight minutes for each revolution. When the next internode is produced the two continue to rotate, and so on with a third. But as the internodes grow old they cease to revolve. It is the terminal two or three of the internodes that exhibit the movement. A shoot was watched when in full rotation with three internodes. The lowest was a little over 8 inches in length, the second was $3\frac{1}{2}$ inches, and the last, or youngest, was 2½ inches. Therefore, the revolving apex of this stem was about 14 inches in length, and it swept round in a circle of 19 inches in diameter, so that the rate of motion must have been but little less than an inch in two minutes and a half, or 23 inches in an hour.

The purpose of this rotation is so self-evident that it scarcely needs explanation. It is undoubtedly directed primarily in search of a support. "This is admirably effected by the revolutions carried on night and day, a wider and wider circle being swept as the shoot increases in length. This movement likewise explains how the plants twine; for when a revolving shoot meets with a support its motion is necessarily arrested at the point of contact, but the free projecting part goes on revolving. As this continues, higher and higher points are brought into contact with the support, and are arrested; and so onwards to the extremity; and thus the shoot winds round its support. When the shoot follows the sun in its revolving course, it winds round the support from right to left, the support being supposed to stand in front of the beholder; when the shoot revolves in an opposite direction, the line of winding is reversed. As each internode loses from age its power of revolving, it likewise loses its power of spirally twining."[1]

---

[1] Darwin, "The Movements of Climbing Plants," p. 15.

By observing the older stems of the hop which have become entwined round a support, or, better still, taking a piece of old hop-bine a few inches in length and splitting it down longitudinally with a knife, we shall observe that the stem is also twisted upon itself; that, in addition to twisting round its support, it twists round its own axis. Mr. Darwin has shown that an internode 9 inches long, and which had revolved thirty-seven times, had become three times twisted round its own axis in the line of the course of the sun. Some have thought in past times that the twisting of the stem caused the rotation, but it is hardly possible that three twists in the stem should have caused thirty-seven rotations. Many twining plants twist in this manner round their own axis. Experiments have demonstrated that when the plant climbs round a smooth support, such as a glass rod, it is much less twisted on itself than when twining round a rough or rugged support. The axial twisting bears some relationship, therefore, to inequalities in the support, as well as to revolving freely without support. Internodes may be observed rotating freely before they have acquired a single twist on their own axis. The axial twisting must, therefore, have some other cause or object than that which has been attributed to it.

We have referred to the wide sweep of the rotating extremity of the hop, but a still more remarkable instance has been recorded in another plant. This was *Ceropegia Gardneri*, a tropical Asclepiad. The top was allowed to grow out almost horizontally to the length of 31 inches. This shoot had three long internodes terminated by three shorter ones. The whole shoot revolved in a direction opposite to the course of the sun, and therefore in a contrary direction to the hop, occupying from five hours and a quarter to six hours and three quarters in each revolution. On account of the great length of this shoot the circle through which it moved was about 5 feet in diameter, or 16 feet in circumference, and the apex travelled through this circle at the rate of 32 or 33 inches per hour. "The weather being hot," Mr. Darwin writes, " the plant was allowed to stand on my study-table; and it was an interesting spectacle to watch the long shoot sweeping this grand circle, night and day, in search of some object round which to twine." [1]

The greater number of twining plants climb in a direction opposed to the course of the sun. The hop and the bryony pursue the same course as the sun. The bean (*Phaseolus vulgaris*), purple convolvulus, and great white convolvulus, follow a course opposed to the sun. The bitter-sweet (*Solanum dulcamara*), which is a poor climber, turns in both directions. Of

---

[1] Darwin, "Movements of Climbing Plants," p. 6.

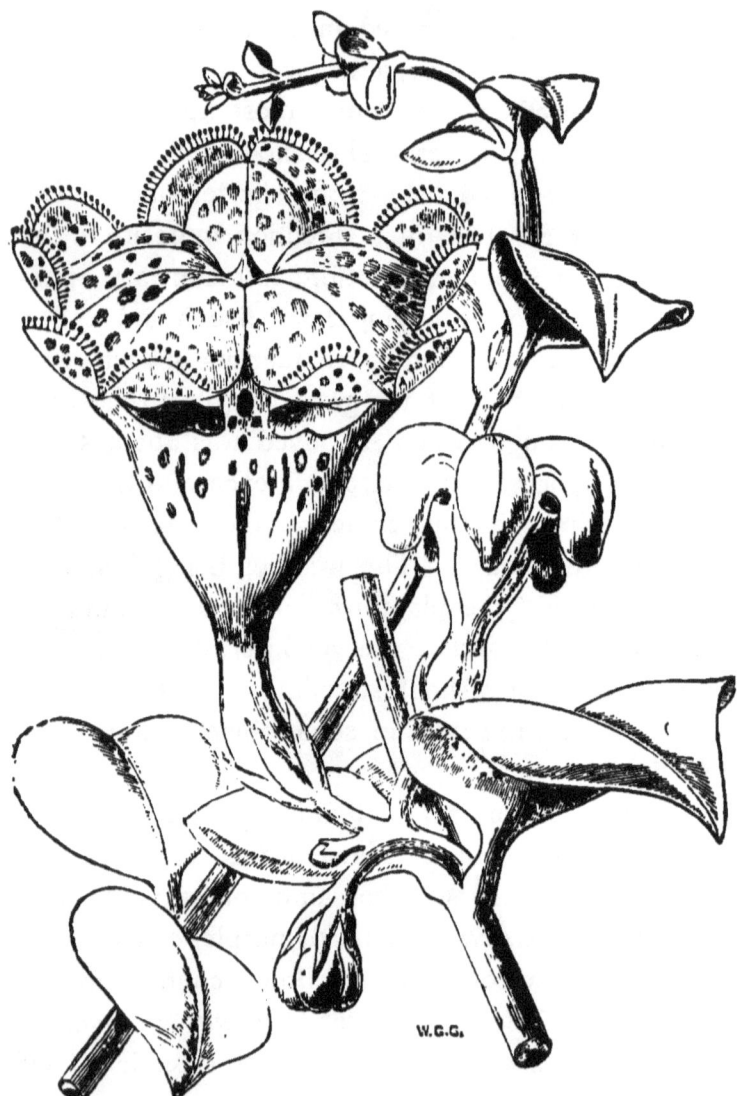

Fig. 22.— Natal Climbing Plant (*Ceropegia Sandersoni*) (from the "Gardener's Chronicle").

the Chili nettle (*Loasa aurantiaca*), a common greenhouse climber, out of eighteen plants eight revolved in opposition to the sun, five followed the course of

Fig. 23.—Bitter-Sweet (*Solanum dulcamara*).

the sun, and four turned first in one direction and then in the other.

During the experiments on twining plants several

interesting facts were evolved which may be mentioned incidentally. For instance, when a revolving shoot is arrested by a stick, and before it has had time to make its first circle round it, the stick is removed, the shoot springs forward, not perhaps to catch the retreating stick, but showing that it must have been pressing against it with some force. After a shoot has wound itself round a stick, if the support be withdrawn, the spiral will remain for a little time, and then the shoot will straighten itself again, and again commence revolving in search of a new support, Although our indigenous twiners are able to ascend by twining round a support as thin as a thread, they cannot twine round an object five or six inches in diameter: the honeysuckle being the only twiner that will encircle trees. Exotic twiners in tropical forests we know will, on the contrary, ascend large forest trees. In all the examples experimented upon the rotation appeared to proceed during the night precisely at the same rate as during the day, showing that light appears to have but little influence on climbing plants. Indeed, one physiologist, Mohl, has affirmed that twining plants are but little sensitive to light.

The conditions under which plants rotate and twine most favourably, are the usual ones under which they would naturally perform the other functions of life, namely, good health and a moderate amount of warmth. The twining Polygonum (*Polygonum convolvulus*), however, twines only during the middle of the summer; in the autumn they will grow vigorously, but without any inclination to climb.

Fig. 24.—The Twining Polygonum (*Polygonum convolvulus*).

Most of the garden beans, of the scarlet runner kind, are excellent twiners, whilst some of the varieties exhibit no tendency to twine.

The only other point to which we shall allude in

twining plants is the different rate of revolution in different plants. In the hop, the shortest period recorded for a revolution was two hours; in the bryony, two hours and a half; in the kidney bean, five minutes less than two hours; the white convolvulus, one hour and forty minutes; in the trumpet-flower (*Tecoma*), six hours and a half; in *Ceropegia*, five hours and a quarter; in a climbing fern (*Lygodeum*), five hours for one species and eight hours for another; in *Lapageria rosea*, eight hours and three-quarters in a hot-house, and eleven hours in a greenhouse; in a species of honeysuckle it was eight hours; and in an exotic (*Sphærostema*) it was eighteen hours and a half.

Although these twiners are described as "those which twine spirally round a support, unaided by any other movement," it has been seen that they possess very remarkable movements of their own, which are intimately related to, and are indeed sufficient to account for, their spiral twining. The same kind of movement, that of rotation, or circumnutation, which we have seen in operation in the young radicle of germinating seeds, in cotyledons, leaves, &c., here reaches a higher development, and achieves a more palpable result.

A remarkable genus of twining plants, belonging to the Amaryllis family, has not yet received the attention they deserve. Passing through one of the "stoves" in Kew Gardens we remarked three or four species of these climbing Amaryllids, and were struck by the peculiar twisting of the petiole of the leaves, which led to their closer examination. In one species (*Bomarea Carderi*) the leaves are lanceolate, on short, flattened foot-stalks. Soon after they are expanded, and before they fall back into their places, the leaves twist over and expose the under surface to the light, so that the true under surface becomes, practically, the upper surface (fig. 25). The most strange circumstance connected with this reversal of the leaves, is the fact that the under surface of the leaf, as though prepared for the twisting, is smooth, and presents the usual characteristic epidermal cells of an *upper* surface, whereas the true upper surface, which by twisting becomes practically the under surface, is furnished with short obtuse hairs, such as might be expected to occur on the under surface of a leaf. In order that

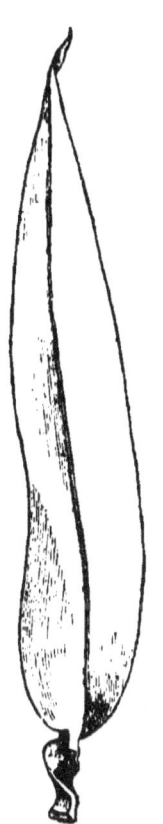

Fig. 25.—Leaf of *Bomarea Carderi*, the petiole twisted in the reversal of the leaf.

there might be no mistake in our interpretation of these facts, we requested Mr. W. S. Gilburt, who has devoted himself successfully to the study of the minute anatomy of plants, to examine and favour us with his opinion. Undoubtedly, he says, the entire structure of the leaf is reversed in order to fulfil the conditions of its reversed position.[1] This seems to us quite an unique illustration of accommodation to circumstances. It still remains to us a puzzle why the leaves should thus reverse themselves. The plants were growing so that the leaves were constantly in contact with small objects, and if the twisting of the petiole was occasioned by effort at clasping, it must have exhibited some evidence, but not a single petiole had embraced anything, and all the leaves had turned over, topside under.

The second class of climbing plants perform this act by means of the ordinary foliaceous organs, or by supplementary ones, which are often modifications of leaves. Those which climb by means of their leaves may do this by embracing the support with the footstalk, or by elongations of the midrib. The most familiar of leaf climbers is the traveller's joy, or clematis, which belongs to a genus including many climbers, such as the splendid large-flowered kinds

---

[1] We are in anticipation that Mr. Gilburt will soon publish the details of his examination of this strange phenomenon.

such great favourites in recent times. Some of the species of clematis retain the power of twining to a limited extent, sometimes in the direction of the sun, and with others in opposition to it. Not uncommonly the same twig will twine two or three times in one direction, then grow erect for a while, and afterwards twine again in the opposite direction. They must therefore be regarded as very inferior twiners. It would be expected, *a priori*, that with this twining power, the terminal joints also rotate, and this is the fact. In one species the quickest revolution was made in five hours and a half, in another in four hours and twenty minutes, in another in three hours and three-quarters, and in another in one hour and fifty minutes.

The petioles, or leafstalks, are so far sensitive to the touch that after being rubbed, or otherwise irritated, they bend towards the point of irritation, and if a stick or twig presents itself in that direction, the leafstalk bends round, and embraces it. If no object is encountered by the bending petiole it soon straightens itself again. The petioles are most sensitive when young; in some species the older petioles lose their power of responding to irritation altogether. In one instance a fragment of thin cotton thread, weighing only one-sixteenth of a grain, caused a petiole to bend perceptibly.

" A thin stick placed so as to press lightly against

a petiole (of *Clematis flammula*) having a leaflet a quarter of an inch in length, caused the petiole to

Fig. 26.—Traveller's Joy (*Clematis vitalba*).

end in three hours and a quarter. In another case

a petiole curled completely round a stick in twelve hours. These petioles were left curled for twenty-four hours, and the sticks were then removed, but they never straightened themselves. I took a twig thinner than the petiole itself, and with it lightly rubbed several petioles four times, up and down; these in an hour and three-quarters became slightly curled; the curvature increased during some hours, and then began to decrease, but after twenty-five hours from the time of rubbing, a vestige of the curvature remained. Some other petioles similarly rubbed twice, that is, once up and once down, became perceptibly curved in about two hours and a half. They became straight again in about twelve hours."[1]

When the petiole embraces a twig it swells perceptibly for two or three days, and ultimately becomes twice as thick as one which has embraced nothing. The same happens also in the case of other leaf-climbers. A section of such a swollen petiole, when examined under the microscope, exhibited an entire change of structure, whereby it had become more rigid and woody, simulating the structure of the stem. It would seem, therefore, that this change in the structure of the clasping petiole is one likely to be serviceable to the plant, by giving greater strength to the curved portion, and thus enabling

---

[1] Darwin, " Movements of Climbing Plants," p. 57.

it to hold more firmly to its support, and withstand greater shocks; in addition to which the greater thickness of the petiole would lessen its chance of being forcibly unwound again from the twig it had embraced. Several species of *Tropæolum* presented somewhat similar phenomena in many respects. They climb also by means of the curvature of the petioles

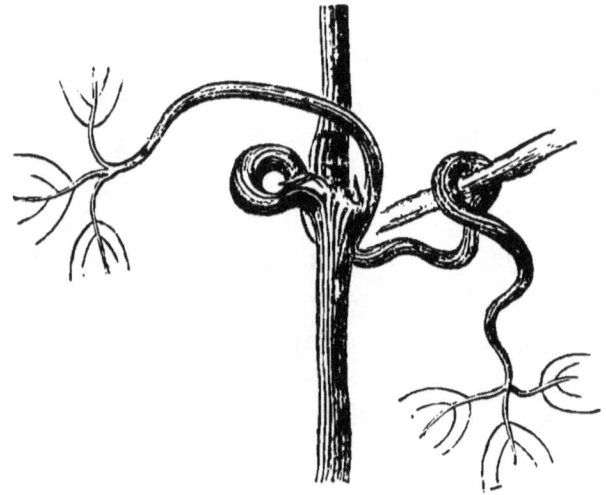

Fig. 27.—Swollen petiole of *Clematis vitalba.*

of the leaves. The petioles are in some species more sensitive than those of clematis. The slightest rub caused them to bend in about three minutes in one case, and in another species the petiole, after a slight rub, became curved in six, eight, ten, and in twenty minutes. It is not unusual to see the green fruit capsules of the common nasturtium in gardens bent over abruptly upon the stem, and even occasionally making a complete turn, or loop. This habit has been noticed also in other species.

Two of the commonly cultivated climbing annuals are leaf-climbers. These are *Maurandia Barclayana* and *Lophospermum scandens*. No special feature necessary to be noted here was developed in the experiments on these plants, but they are mentioned chiefly on account of the facility with which they may be cultivated by those who may desire to repeat these observations for themselves, and trace all the phenomena of leaf-climbing.

The little fumitory (*Fumaria officinalis*) is also a humble example of a climber of this kind (fig. 28). Some of the petioles were determined to be sensitive to touching, and responded thereto in about an hour and a quarter. The young internodes forming the terminal shoots of the stem and branches are in constant rotation. The leaves also have their own special spontaneous movement. As this plant is a common weed there need be no difficulty in verifying, and even supplementing, the observations already made. The *Corydalis* is a closely allied plant, but not so common; it is intermediate between leaf-climbers and tendril-bearers, with some of the habits of both (fig. 29).

The plants which climb by means of the development of the tips of the leaves into hooks, are so few, and those are exotic, that we may dismiss them with a brief explanation of the process. The end of the leaf (in *Gloriosa Plantii*) forms a narrow projection, which is thickened and at first nearly straight; subsequently it bends downwards and forms a hook, which becomes strong enough and rigid enough

Fig. 28.—Common Fumitory (*Fumaria officinalis*).

to catch any object and fasten the plant. The inner surface of the hook is somewhat sensitive, and, when a twig is caught by it, the extremity curves a little inwards and permanently seizes it. If nothing is caught the hook remains open and sensitive for some time, but ultimately the extremity slowly curls inwards and forms a coil at the end of the leaf. In one leaf the hook remained open for thirty-three days. When the tip has curled into the form of a

Fig. 29.—Climbing Corydalis (*Corydalis claviculata*).

ring all sensibility is lost, but as long as it remains open some sensibility is retained.[1]

We now pass on to tendril-bearers, premising that tendrils are in most cases modifications of leaves

---

[1] Darwin, " Movements of Climbing Plants," p. 79.

transformed into filaments, which are used wholly for climbing. In other words, a tendril may be a leaf so modified that it is reduced to the midrib and a few lateral branches, with none of the functions of leaves, but with a new and special function contemporaneous with the modification, viz., that of enabling the plant to climb and maintaining it in that position. But a tendril may also be a modification of the flower-stalk, or of some other organ. It matters not, in so far as the present inquiry is concerned, what organs are so modified; in fact, botanists themselves do not seem to be entirely agreed on this point.

Very few plants with tendrils possess the power of climbing up an erect stick, but most of them exhibit rotation in the growing points, performing revolutions not unlike in character to those of twiners, and in like manner in different directions. This movement, though similar in its action, has a different purpose. In twiners the oscillation is evidently in search of some object around which to entwine; in tendril-bearers in order to bring the tendrils in contact with some support. The tendrils themselves also rotate in many species; in some the tendrils, internodes, and petioles, move in harmony together. In *Cobœa scandens*, a well-known climber in common cultivation, the tendrils are ten or eleven inches in length, and revolve rapidly and vigorously. Three large circular sweeps were observed within an hour and a quarter, but the growing point does not rotate. In *Echinocystis lobata,* a plant of the cucumber family, the tendrils, which are from seven to nine inches in length, revolve as well as the internodes, but over a wider surface. The circles swept by the tendrils are from fifteen to sixteen inches in diameter, whilst those of the internodes are not more than about three inches. The quickest rate of motion for the completion of a revolution was about one hour and three-quarters.[1] In a passion-flower the internodes as well as the tendrils rotate, the former very rapidly, performing its revolution in an average period of about an hour. In a species of trumpet-flower (*Bignonia littoralis*) the mature tendrils rotate much slower than the internodes, the former taking six hours to perform a revolution, and the latter two hours and three-quarters. In the Virginia creeper neither the internodes nor the tendrils possess the power of rotation.

That tendrils are sensitive to a touch, one might expect from the purposes they are called upon to serve, but this faculty varies in different species. In one of the passion-flowers (*Passiflora gracilis*) where the tendrils are thin, delicate, and straight, except the curved tips, a single delicate touch on the concave

---

[1] Darwin, "Movement of Climbing Plants," p. 128.

surface of the tip caused it to curve immediately, so that in two minutes it formed an open spire. The movement was generally perceptible within half a minute after being touched. A tendril which curls through being touched, but does not embrace anything, straightens itself again, but soon becomes irritated by a second touch. In order to ascertain how often the same tendril may be excited one tendril was selected, and this alternately straightening itself, answered to the stimulus no less than twenty-one times in fifty-four hours.

Professor Asa Gray has observed an equally rapid response to a touch in the tendrils of a plant of the cucumber family, but instances of such rapidity are rare. In some, the movement takes place after a few minutes, in others it is an hour or two, but in all some exhibition of sensibility has been observed. It is noteworthy that drops of water sprinkled with a syringe, so as to resemble falling rain, in no instance appeared to have the least stimulating effect. In most cases a touch from another tendril seemed to have no influence, although, in the bryony and the vine other tendrils have been seen embraced.

The sensibility of tendrils to light may also be illustrated by the trumpet-flower (*Bignonia capreolata*). In his experiments on these plants, Mr. Darwin observes, " In two instances, a pair of leaves stood so that one of the two tendrils was directed towards the light, and the other to the darkest side of the house; the latter did not move, but the opposite one bent itself first upwards and then right over its fellow, so that the two became parallel, one above the other, both pointing to the dark. I then turned the plant half-round, and the tendril which had turned recovered its original position, and the opposite one which had not before moved, now turned over to the dark side. On another plant, three pairs of tendrils were produced at the same time by three shoots, and all happened to be differently directed. I placed the pot in a box open only on one side, and obliquely facing the light. In two days all six tendrils pointed with unerring truth to the darkest corner of the box, though to do this each had to bend in a different manner. Six wind-vanes could not have more truly shown the direction of the wind than did these branched tendrils the course of the stream of light which entered the box. I left these tendrils undisturbed for about twenty-four hours, and then turned the pot half-round; but they had now lost their power of movement, and could not any longer avoid the light."[1] The rotation in the tendrils of some plants is retarded and in others accelerated by the action of

---

[1] Darwin, "Movements of Climbing Plants," p. 98.

P

the light. Those of the pea, and some others, seem to be insensible to its influence.

The mode by which tendrils clasp and attach themselves to their supports is variable, even in the same genus. In some, they twine spirally, like a corkscrew; in others they grasp a projection in a manner resembling the foot of a bird; in others, again, they attach themselves by hooks or grapnels; and in others, the sharp points are inserted in cracks and fissures, or minute holes, although this latter in some cases, seems to be only a temporary expedient. The most elaborate mode of attachment is one in which the tips of the tendrils undergo special modification, and to this kind we must advert more in detail.

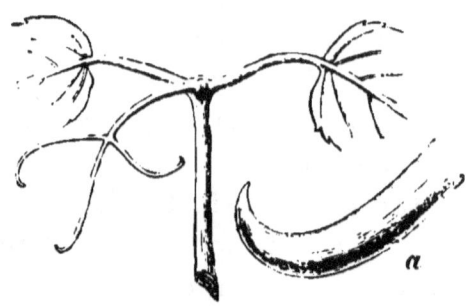

Fig. 30.—Hooked tendril, like foot of a bird, from *Bignonia Tweediana*. Tip of hook magnified (*a*).

This curious but interesting adaptation of the tendrils of a plant, in order the better to fulfil its function of climbing, is related of an exotic trumpet-flower (*Bignonia capreolata*). The tendrils are branched, having about five branches, each of which is divided again at the apex, with each point blunt but distinctly hooked. Having placed a piece of wood containing numerous cracks within reach of the plant, it was observed that the tips of the immature tendrils crawled like roots into the minutest crevices. In two or three days after the tips had thus crawled into the crevices, or after the hooked ends had seized on projecting points, another process commenced. The tips of the inner surfaces of the hooks begin to swell, and in two or three days are visibly enlarged. After a few more days the hooks are converted into whitish balls, rather more than the one-twentieth of an inch in diameter, and composed of coarse cellular tissue, sometimes enveloping and concealing the hooks themselves. The surface of the balls secrete a viscid matter, to which small objects adhere. When slender fibres become attached to the balls the tissue grows round and over them, and fresh fibres continuing to adhere, as many as fifty or sixty fibres of flax have been counted imbedded in one of these balls. The fibres are clasped so tightly that they cannot be withdrawn.[1] When two balls from adjacent extremities come into contact they will sometimes coalesce. If the hooked extremities of the tendrils do not touch anything the discs are not formed in this species, although, in an allied plant, Fritz Muller has remarked that smooth shining discs

---

[1] Darwin, "Movements of Climbing Plants," p. 101

terminate the tendrils without their having come into contact with any object.[1]

The Virginia creeper (*Ampelopsis hederacea*) has also branched tendrils five or six inches in length. The tips of the branches are at first curved, and when they come in contact with a wall, or other flat

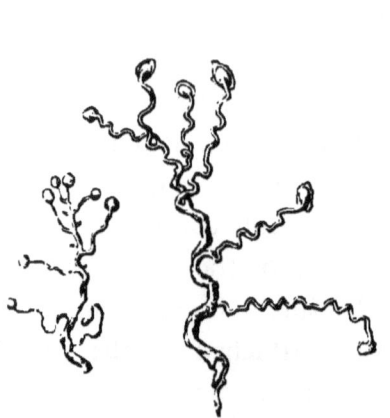
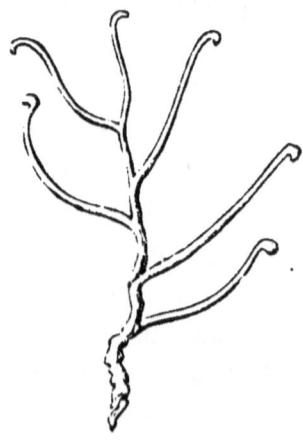

Fig. 31.—Tendrils of Virginia creeper, with discs attached.

Fig. 32.—Tendrils of Virginia creeper, discs not attached.

surface, the hooks are brought into apposition to it. In the course of two days after a tendril has arranged its tips so that they touch and press on the surface, the curvatures swell, become bright red, and form little discs or cushions on the under side. In one case the tips were swollen in 38 hours, and in another 48 hours, and in an additional 24 hours were firmly

---

[1] Muller, "Journal of Linnæan Society," ix., p. 348.

attached to a smooth board. The discs are generally formed on one side of the curved tip, and never, as far as yet observed, without coming in contact with some object.[1] Dr. McNab[2] has observed in another species that small globose discs are formed before the tips come into contact. This also corresponds with the observations on Bignonia.

It seems evident that these discs possess the power of secreting some resinous cement, by means of which they adhere to the support to which they attach themselves. When a tendril does not become attached, its primary object being frustrated, in the course of a few weeks it shrinks and withers, and finally drops off. When the discs have become attached, then the tendril contracts spirally, so as to become very elastic, and at the same time thickens so as to attain increased strength. Even after the tendrils are dead they still continue to adhere, and retain strength. One single branch of a tendril, which had been dead at least for ten years, still remained elastic, and capable of supporting a weight of two pounds, so that assuming all the branches of the same tendril to have been equally attached, and equally strong, the entire tendril would be capable of enduring a strain of ten pounds. Sachs

---

[1] Darwin, "Movements of Climbing Plants," p. 145.
[2] Dr. McNab, in "Transactions of Botanical Society, Edinburgh," xi., p. 292.

remarks, that the tendrils of different species are adapted to clasp supports of different thicknesses,

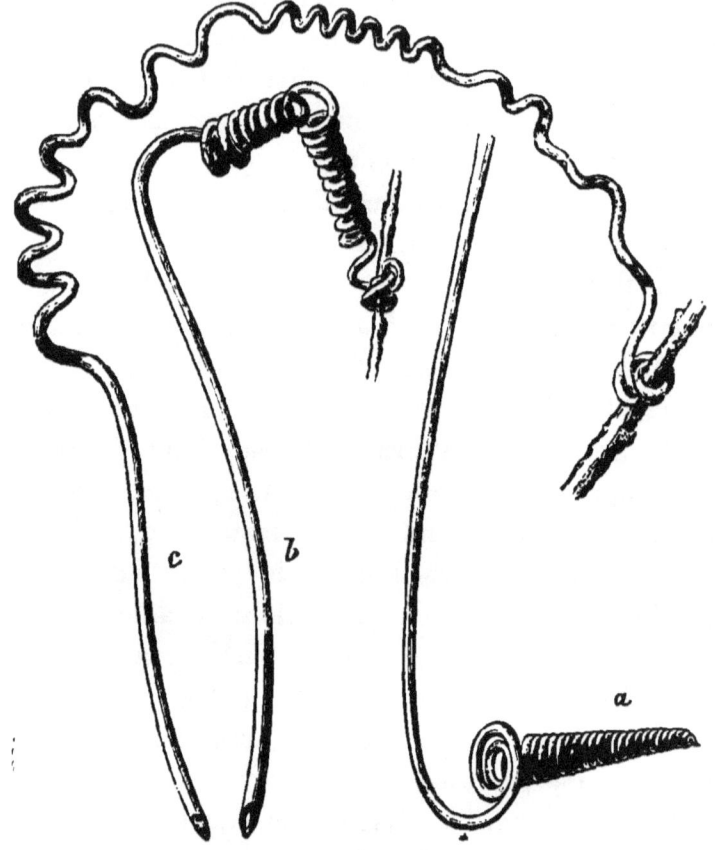

Fig. 33.—Tendrils of *Passiflora edulis*.

and that when a tendril has clasped its support it afterwards tightens its hold.[1]

When a tendril does not attach itself it ultimately

---

[1] Sachs' "Text-book of Botany," p. 280.

winds up into a close spiral (fig. 33, *a*), but if it attaches its extremity to any object it winds itself into a more open spiral for some distance, then reverses, and winds in the opposite direction (fig. 33, *b, c*). The reason for this will be obvious if we attempt to twist a piece of twine with its extremity fixed; the torsion will soon become so great that we

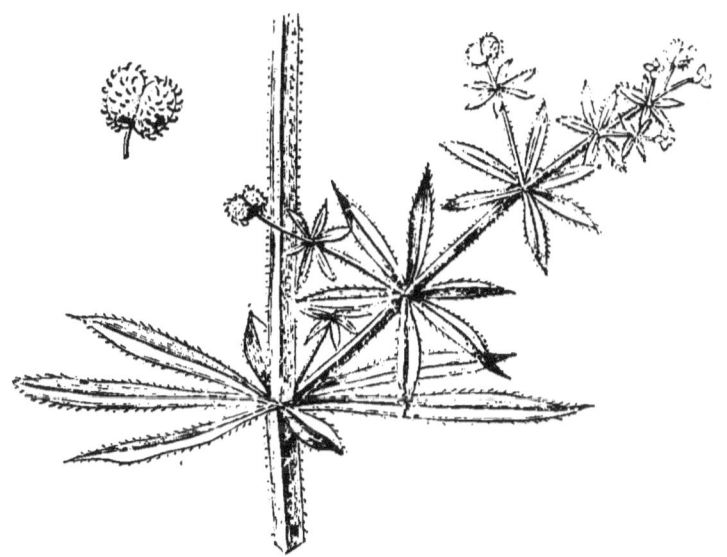

Fig. 34.—Cleavers (*Galium aparine*).

must cease or reverse the spiral. The latter movement relieves the torsion, and the twist in the second direction soon compensates the first. If any tendril with its extremity attached be examined, this reversal of the twist will be found of universal occurrence. Indeed, it must be so, as a physical necessity, to which the tendril is compelled to submit. The above figures in illustration are from a cultivated passion-flower (*Passiflora edulis*).

There remain only the two sections, of scramblers, or plants which ascend merely by hooks, and root climbers, which ascend by means of rootlets, to be described. As these do not exhibit many remarkable phenomena a few observations will suffice. The scramblers are represented by that very common weed the "cleavers" or "goosegrass" (*Galium aparine*), which scrambles up hedges and amongst thickets by means of the recurved hooks with which the stems are liberally provided. The young shoots appear to possess no spontaneous rotation, and the climbing habit is literally reduced to a scrambling, the lowest and most imperfect climbing with which we are acquainted. Some kinds of roses would also find a place in this section, for they will scramble up the walls of a house if there is a trellis-work to assist them. Professor Asa Gray,[1] explaining this phenomenon, in reference especially to the Michigan rose (*Rosa setigera*), remarks that the summer shoots are strongly disposed to push into dark crevices and away from the light, so that in pursuance of this habit they would be sure to thrust themselves under a trellis, whilst the lateral shoots, developed in the following spring, will emerge from the trellis in search of the light. This alternate mode of growing

---

"American Journal of Science," vol. xl., p. 282.

inwards and outwards is just the process which would be mechanically adopted to secure a rose to a trellis-work.

Of root-climbers our most familiar indigenous illustration is the ivy (*Hedera helix*), which ascends by means of rootlets, which adhere to the wall or old trunks, and thus enable the plant to reach the summit of its ambition. Dr. Spruce, alluding to a South American plant (*Marcgravia umbellata*) which grows against the trunks of trees by means of claspers or roots, remarks, that when it has reached the light and the branches become free, the stems which before were flattened become rounded, and the leaves are altered in character and general appearance. To a certain extent this is also true of the ivy, for when it has reached the top of an old trunk and the free branches are produced, they are destitute of rootlets, and the leaves are smaller, more narrowed towards the footstalk and otherwise modified.

There is also a species of fig (*Ficus repens*) which climbs a wall in the same manner as the ivy. The rootlets of this plant, when pressed lightly on slips of glass, were found to emit minute drops of a clear fluid, which is slightly viscid. This fluid exhibited the remarkable faculty of remaining fluid during 128 days. Other rootlets, left in contact with glass for a longer period, secreted larger drops of fluid, which were more tenacious, and could be drawn out in threads. Other rootlets, left for a still longer period in contact with glass, became firmly cemented to it, and when torn away atoms of yellowish matter were left behind. The inference from these observations, strengthened by chemical tests applied to the secretion, is that the fig has the power of transuding from the rootlets a kind of cement, similar to caoutchouc, by means of which the rootlets become attached to the supporting object.[1]

As we have intimated, the tendril-bearers seem to be the most highly organised of climbing plants. The most interesting point in their history is, as Mr. Darwin has pointed out, the varied movements they display according to their wants. The first action of a tendril is to place itself in a proper position. Secondly, if a twining plant, or tendril, gets into an inclined position accidentally it soon bends upwards again. Thirdly, climbing plants bend towards the light by a movement analogous to that which causes them to revolve so that their revolution is accelerated or retarded in travelling to or from the light. A few tendrils bend towards the dark. Fourthly, there is the spontaneous rotation which is independent of external stimulus. Fifthly, tendrils all have the power of movement when touched, and bend towards the point of irritation. If the pressure be not permanent, the part soon straightens itself again. Lastly, the tendrils soon after clasping their support, effectually contract themselves in a spiral manner.

---

[1] Darwin. "Movements of Climbing Plants," p. 186.

Reflecting upon these movements we are prepared to assent to the concluding paragraph of the work in which most of the observations in this chapter have been founded. " It has often been vaguely asserted that plants are distinguished from animals by not having the power of movement. It should rather be said that plants acquire and display this power only when it is of advantage to them ; this being of comparatively rare occurrence, as they are affixed to the ground, and food is brought to them by the air and rain. We see how high in the scale of organization a plant may rise, when we look at one of the more perfect tendril-bearers. It first places its tendrils ready for action, as a polypus places its tentacula. If the tendrils be displaced it is acted on by the force of gravity and rights itself. It is acted on by the light and bends towards or from it, or disregards it, whichever may be most advantageous. During several days the tendrils or internodes, or both, spontaneously revolve with a steady motion. The tendril strikes some object, and quickly curls round and firmly grasps it. In the course of some hours it contracts into a spire, dragging up the stem and forming an excellent spring. All movements now cease. By growth the tissues soon become wonderfully strong and durable. The tendril has done its work, and has done it in an admirable manner."[1]

---

[1] Darwin, " Movements of Climbing Plants," p. 206.

CHAPTER X.

SENSITIVE PLANTS.

CULTIVATED in green-houses as curiosities several species of exotic plants have received the name of "sensitive plants." These are, perhaps, the most decided in their exhibition of irritability, or movement, when touched; but the same phenomenon in a less degree is to be found in a vast number of plants. Poets have taken advantage of this extraordinary faculty, and invested those which possessed it with mystery and romance.

> A sensitive plant in a garden grew,
> And the young winds fed it with silver dew,
> And it opened its fan-like leaves to the light
> And closed them beneath the kisses of night.

Travellers in foreign climes have delighted to descant on the wonderful sensitive plants. "Looked upon with such interest in our green-houses, but which here abound (Brazil) as common as wayside weeds. Most of them have purple or white globular heads of flowers. Some are very sensitive, a gentle touch causing many leaves to drop and fold up; others require a ruder hand to make them exhibit their peculiar properties, while others, again, will scarcely show any signs of feeling, though ever so roughly treated. They are all more or less armed with sharp prickles, which may partly answer the purpose of guarding their delicate frames from some of the numerous shocks they would otherwise receive."[1]

One of the best known "sensitive plants" is the one usually called the "sensitive plant"—that is, the *Mimosa pudica* of botanists, a plant of which stands on the table before us as we write. In this the leaves are bipinnate, then quadripinnate. There is a pair of pinnæ at the end of a long peduncle; with maturity two others are developed. These pinnæ consist each of about eight to twelve pairs of opposite leaflets, the two pinnæ standing almost at right angles to each other. At a slight touch all the leaflets rise and close the upper surfaces together, at the same time the two pinnæ approach each other so as to be nearly parallel, instead of at right angles as before. In this manner the leaves which have been touched respond, and remain closed for some time; but at length they recover gradually from the shock, and return again to their previously expanded position. The experiment may be repeated with similar results; but if repeated again and again the movements

---

[1] Wallace, "Travels on the Amazon," p. 11.

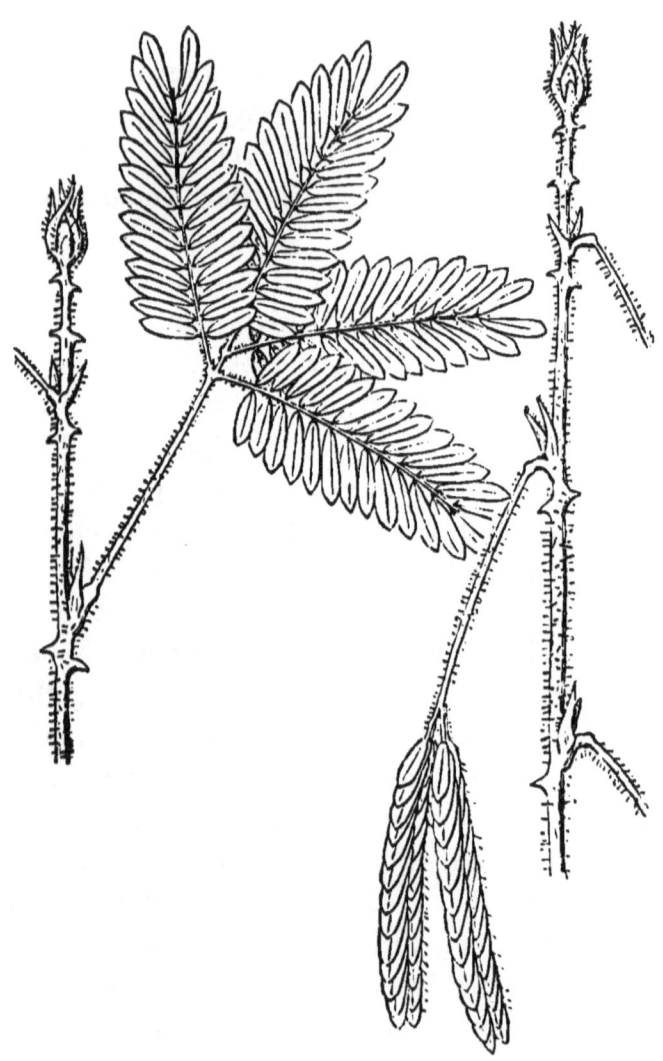

Fig. 35.—Leaves of sensitive plant, *Mimosa pudica*, awake and asleep.

become more tardy, as if debility ensued from over-exertion. Too strong sunlight has a similar effect in causing the leaflets to close. A strong puff of the breath, or a shake of the pot, is enough to cause the movement. It is by no means a slow and gradual change, but an almost instantaneous one, sometimes of the leaflets on both sides simultaneously, and sometimes first of one side and then the other. The return movement is much more deliberate, so that it can scarcely be detected.

Yet more remarkable movement takes place in another celebrated plant, without a touch being required to stimulate it. This is the "telegraph plant" (or *Desmodium gyrans*), a native of Bengal. The lateral leaflets keep constantly moving all day long without any external impulse being given to them. They move up and down and circularly, this last motion being performed by the twisting of the footstalks, and while one leaflet is rising its corresponding one opposite is generally being depressed. The motion downwards is generally quicker, or more irregular than the motion upwards, which is steady and uniform. These motions are observable for twenty-four hours in the leaves of a shoot which is lopped off from the plant, if kept in water. If from any obstacle the motion is retarded, upon its removal it is renewed with greater velocity. The motion is most evident when the sun's rays are upon the plant.

This shrub[1] belongs to the same natural order as the acacia, the furze, and the broom. Another, which belongs to the same order as our little wood-sorrel, but, on the contrary, grows to a big tree, is also remarkably sensitive. It is the camrunga-tree of India (*Averrhoa carambola*). The leaves are pinnated, or feathered, with alternate leaflets, and an odd one at the end. Their common position in the daytime is horizontal. On being touched they move downwards, frequently in so great a degree that the two opposite leaves almost touch one another by their undersides, and the leaflets sometimes either come into contact, or even pass each other. The whole of the leaflets of one leaf move by striking the branch with the finger-nail, or each leaflet can be moved singly by making an impression which shall not extend beyond it. Thus the leaflets of one side of the leaf may be made to move one after another, whilst the opposite ones continue as they were, or they may be made to move alternately in any order by merely touching the leaflet intended to be put in motion. After sunset the leaves go to sleep, first moving down so as to touch one another by their undersides; they, therefore, perform a greater motion at night of themselves than they can be made to do during the day by external impressions. The

---

[1] Hogg's "Vegetable Kingdom."

rays of the sun may be concentrated by a lens upon the leaflets without producing any motion; but when directed upon the leafstalk the response is almost instantaneous. The leaves move rapidly under the influence of an electric shock.[1]

> The sensitive plant was the earliest
> Up-gathered into the bosom of rest;
> A sweet child weary of its delight,
> The feeblest and yet the favourite,
> Cradled within the embrace of night.[2]

The leaves of the common wood-sorrel not only close in the evening, but if gathered roughly seem to shrink from the touch like the sensitive plant. Another species of wood-sorrel (*Oxalis sensitiva*), which is a native of Amboyna, is said to be so delicately sensitive that it will not bear the blowing of the wind upon it, without contracting its leaves. Dr. Roxburgh said of it, that "it was like a maiden, though common on every wayside, it may be looked at but is not to be touched."

Thus much is sufficient to explain what are the most manifest phenomena of those which have received the name of "sensitive plants": in other words, plants which exhibit irritability in a manifest degree.[3] In these instances the leaves possess the

---

[1] Hogg's "Vegetable Kingdom." [2] Shelley.
[3] Of plants which are reputed to possess this power, the following may be named:—*Desmodium gyrans, Mimosa pudica,*

power of moving themselves in response to a touch, either by elevation or depression, such movement being independent of another motion, termed the "sleep of plants," which is exhibited as daylight declines and night comes on, by the gradual folding or closing of leaves. There is undoubtedly an intimate relationship between the phenomena of motion in leaves when touched, and in those which close spontaneously on the decline of light, and also in such plants as turn themselves towards or away from the sun. Nevertheless, for convenience, we have preferred to write separately of "sensitive plants," of the "sleep of plants," and "heliotropes," or "sun-turners."

It is needless to explain that sensibility, as implied in the term "sensitive plants," does not exist in the vegetable kingdom in the same manner as in the animal. Without brain, and without nervous system, that which we characterize as *sensibility* does not exist. Yet there is an apparent sensibility to external impressions, and there is also the power of transmitting impressions from one part of the plant to the other. Who will attempt, and how is the limit to be defined to sensibility, or what

---

M. sensitiva, M. casta, Æschynomene sensitiva, Æ. indica, Æ. pumila, Smithia sensitiva, Desmanthus stolonifer, D. triqueter, D. lacustris, Oxalis sensitiva, and, in a small degree, *O. stricta, O. acetosella, O. corniculata* and *O. Deppei.*

other term have we which will adequately supply its place, when applied to plants? A curious coincidence of the effects produced on animals and sensitive plants, under the same conditions, was demonstrated by M. Blondeau. Induction currents of electricity have little or no effect on animals when under the influence of anæsthetic agents. In order to see what would be the effect, in the case of the *Mimosa pudica* under like circumstances, he exposed a specimen to the anæsthetic effect of a few drops of ether sprinkled in the glass enclosing the plant. In a short time the plant experienced the effect of the anæsthetic, its leaves refused to move when shaken, and manifested no sensibility even when the induction current was passed through them.[1] The same experimenter found that when a current from a galvanic battery was passed through the leaves of this "sensitive plant," no result was produced, and the plant did not respond to the stimulus. On the other hand, when in place of the direct an indirect current was employed, by the use of a small Ruhmkorff's coil, the results were entirely different, the leaflets folded up, and the leafstalks drooped along the whole course of the stem. If the current were continued for a short time the plant after a period of repose raised its leaves, and resumed its ordinary state; but if the

---

[1] "Popular Science Review," vii., p. 32.

experiment was prolonged for twenty-five minutes, the organism seemed to become entirely exhausted, and the following day was found withered and blackened, as though struck by lightning.

Some interesting observations have been made by Dr. Maxwell Masters on the effects of ether upon plants, which have a relationship to this phase of the subject. He writes:[1] "On allowing a drop of ether to fall on one of the leaflets of *Mimosa pudica* from a height of five or six inches, contraction of the leaflet instantly took place, and was immediately followed by the motion, in successive order, of the adjacent folioles, proceeding from the apex towards the stalk of the leaf. When, on the other hand, the drop of ether was placed as gently as possible on the surface, the leaflet did not move, but seemed paralysed by the anæsthetic agent, while the adjacent ones, not touched by the ether, moved as in the preceding case. Ether spray applied with the jet had precisely similar effect. When the spray fell directly on the leaflets, that is, with some force, the impact of the falling drops counteracted any paralysing power that the ether might have; but when the spray was so directed as not to fall directly or with force on the leaflets, then such of them as came within its influence were rendered motionless, the adjacent folioles

---

[1] "Popular Science Review," vii., p. 30.

contracting from the apex towards the base as before. A spray of water directed on to the leaflets caused them to fall, but if not allowed to impinge directly on them no motion ensued, though, of course, the water did not, as the ether did, stop their mobility, as a touch was sufficient to make them collapse after the water spray, while after the ether spray contact produced no effect."

"The effect of the ether spray on certain other plants was, in two instances, remarkable, though the results now to be mentioned were only obtained in two instances out of many trials on various plants in hot-houses in November, 1867. On applying the spray to the extremity of one leaf of *Iresine Herbstii*, which from having been grown in heat, was what gardeners call 'drawn,' that is, had comparatively long intervals between the leaves, and a flaccid texture, a thin film of ice was speedily produced on the distal end of the leaf. In less than two minutes the whole shoot, four or five inches long, was observed to bend quickly downwards. Next morning the whole shoot was dead. To what precise circumstances the rapid transmission of the effect from one end of the shoot to the other, and its ultimate death, are due, it would be premature to assert, as it is difficult in such a case to eliminate the irritant effect of the ether (clearly it did not here act as an anæsthetic) from the effect of the cold, and ice produced by its rapid evaporation. It may be stated, that two or three drops placed on the leaf in the ordinary way had no effect at all. A few days after, similar trials were made in the Botanic Garden, Chelsea, on some plants of the same species, grown in a colder house, and which were 'short-jointed,' and altogether firmer in texture. In these instances no other effect was produced then the death of the leaf. The other case was a Maranta, also growing in a stove, and in which the application of ether spray to the tip of a leaf caused it to roll up on to the underside, like a roll of paper. In the young state the leaves of this plant are rolled lengthwise, but the effect of the ether was to cause the leaf to roll up along the under surface, from the tip towards the stalk."

Returning from this digression to the special subject of this chapter, it may be intimated that temperature and a healthy condition are important factors in the manifestations of irritability. The tropical wood-sorrel (*Oxalis sensitiva*), which is remarkably sensitive in countries where it grows naturally, scarcely exhibits this peculiarity, even when grown in hot-houses, in this country. Both *Desmodium gyrans* and *Mimosa pudica* show an evident increase of susceptibility with an elevation of temperature. A plant of the Mimosa, carried about in his carriage by Desfontaines, although at first exhibiting its usual susceptibility, by and by ceased to respond to the stimulus, and its leaves became motionless as a result of the continued vibration. Dr. Masters experienced similar results in a plant carried by railway. It is stated that the Mimosa, which is an annual, is more feeble towards the close of the year than when in full vigour. In its native country, where the plant grows wild, it is said that the touch of children affects the plant's movements more than that of adults will do; and Dr. Sigerson has stated that it is more active in its movements when excited by a person in a tonic condition than when he is weary or exhausted. These are curious facts, although they contribute but little to the solution of the problem. "Why a touch," writes Mr. Darwin, "slight pressure, or any other irritant, such as electricity, heat, or the absorption of animal matter should modify the turgescence of the affected cells in such a manner as to cause movement, we do not know. But a touch acts in this manner so often, and on such widely distinct plants, that the tendency seems to be a very general one; and, if beneficial, it might be increased to any extent."[1] We may, therefore, be excused from any attempt to explain that which such an experienced authority confesses that "we do not know." The

---

[1] "Movements of Plants," p. 571.

facts themselves are placed on record, and hereafter the key to the mystery may perhaps be found.

Spontaneous movements are common in the organs of reproduction, being more or less associated with the process of fertilization. "In Stylidium, an Australian genus, the style and filaments are adherent into a column, which hangs over on side of the flower. When touched it rises up and springs over to the opposite side, at the same time opening its anthers and scattering the pollen. The stamens of the various species of berberry (*Berberis* and *Mahonia*) exhibit this irritability to a remarkable degree. If touched with a pin or other object at the base of the inside face of the filament, the stamen will spring violently forward from its place within the petal, so as to bring the anther into contact with the stigma, and will, after a time, slowly resume its original position. At first sight it may seem as if this contrivance were intended to ensure the fertilization of the pistil from the pollen of its own flower. In reality, however, the reverse is the case; the excitation takes place in nature when an insect entering the flower for the sake of the honey in the glands at the base of the pistil touches the inside of one of the stamens. The pollen is thus thrown on to the head or body of the insect, which carries it away to the next flower it visits, and leaves some of it on the stigma, and thus cross-fertilization instead of self-fertilization is secured.

Similar motion of the stamens towards the pistil, but spontaneous, takes place in the London Pride and other species of Saxifraga."[1] This may be studied also in that pretty marsh-flower, the grass of Parnassus (*Parnassia palustris*). In this flower each of the white petals has a glandular appendage at the base which forms a portion of the disc. It consists of a crescent-shaped scale, bearing on its margin from sixteen to twenty slender hair-like pedicels, each of which supports a globose vesicle filled with fluid. In the centre of the flower is a rather large

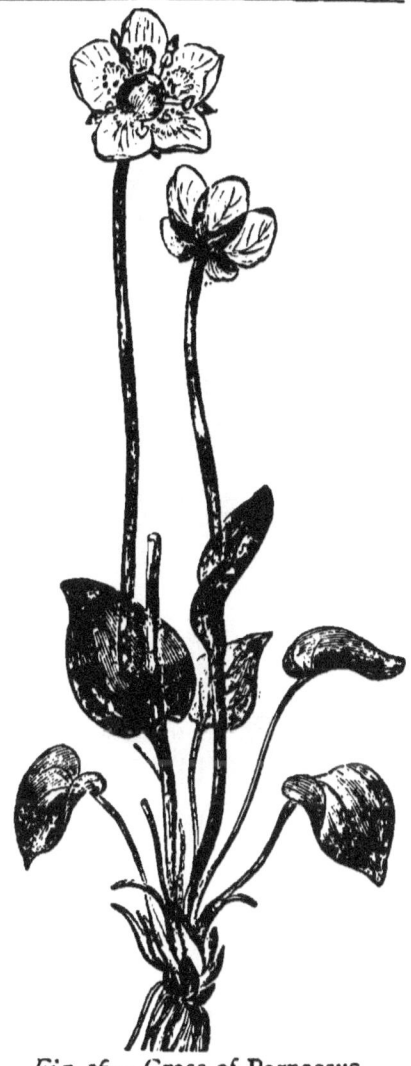

Fig. 36.—Grass of Parnassus (*Parnassia palustris*).

---

[1] A. W. Bennett, in "Popular Science Review," vol. xi. (1872), p. 378.

ovary, with a stigma on the top, and surrounding this five stamens on short filaments, which are indeed so short that the anthers can scarce reach the top of the ovary. When the pollen is ripe the filaments of the stamens lengthen themselves one by one, and apply the face of the anther to the stigma. After the pollen is

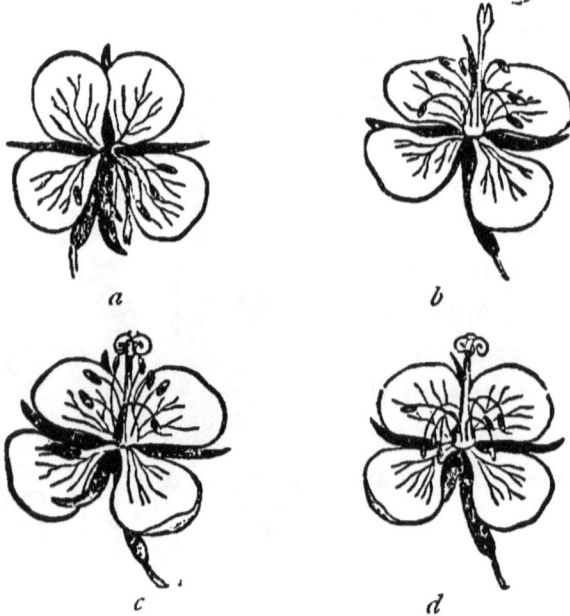

Fig. 37.—Flowers of Epilobium.

discharged the filaments bend back to the level of the petals, and the empty pollen-cases soon fall off and leave the five filaments spread out in a star-like manner alternate with the petals of the flower. The same phenomena have been observed in other species of the same genus.

Similar movements may be observed in the rose-bay willow herb (*Epilobium angustifolium*). The pistil in the centre of the flower has a four-lobed stigma, supported on an erect style, which is rather longer than the filaments of the stamens. When the flower first opens, the lobes are closely applied to each other, and the style and stamens hang down (*a*). As the anthers become mature the style becomes erect, and the stamens begin to elevate themselves (*b*). By the time the anthers are fully matured the lobes of the stigma divide and curl outwards and downwards in a circinate manner, so that they may be reached by the anthers; the filaments become erect, and the pollen is discharged upon the lobes of the stigma (*c*). After discharging the contents of their anthers the stamens droop and become pendulous again, whilst the style remains erect (*d*).

Peculiar movements have been observed in other parts of flowers, and in some orchids, as in *Megaclinium falcatum*, the lip, or labellum, is said to exhibit spontaneous movement. Alluding to these remarkable plants, Dr. Lindley says,[1] "Among many other remarkable peculiarities the irritability of the labellum must not be passed over in silence. This is extremely striking in some species."[2] In *Calcana nigrita* the

---

[1] "Vegetable Kingdom," p. 179.
[2] As in various species of *Pterostylis*, in the genus *Megaclinium*, and in *Bolbophyllum barbigerum* and *Careyanum*.

column is a boat-shaped box, resembling a lower lip; the labellum forms a lip that exactly fits it, and is hinged on a claw which reaches the middle of the column. When the flower opens, the labellum turns round within the column and falls back, so that, the flower being inverted, it stands fairly over the latter. The moment a small insect touches its point, the labellum makes a sudden revolution, brings the point to the bottom of the column, passing the anther in its way, and thus makes prisoner any insect which the box will hold. When it catches an insect it remains shut while its prey continues to move about, but if no capture is made the lid soon recovers its position. Another plant (*Drakœa elastica*) has a single flower placed at the end of a slender, smooth, erect scape, from twelve to eighteen inches high; and its labellum, which is hammer-shaped, and placed on a long arm, with a movable elbow-joint in the middle, is stated to resemble an insect suspended in the air, and moving with every breeze. Another plant of this description is *Spiculœa ciliata*, whose rusty flowers when spread open may be compared to long-legged spiders; the lip, with a long solid lamina, looking like their body, while an appendage at its apex, which is apparently movable, is not unlike the head of such a creature."

One of the species with an irritable labellum is called the Toad Orchis (*Megaclinium bufo*), and was figured and thus described, many years ago, in the "Botanical Register." "Let the reader imagine a green snake to be pressed flat, like a dried flower, and then to have a row of toads, or some such speckled reptiles, drawn up along the middle in single file, their backs set up, their fore-legs sprawling right and left, and their mouths wide open, with a large purple tongue wagging about convulsively, and a pretty considerable approach will be gained to an idea of this plant, which, if Pythagoras had but known of it, would have rendered all arguments about the transmigration of souls superfluous."[1] This orchid is a native of Sierra Leone.

Mr. Darwin, writing in general terms, says, "Of the many singular properties of orchids the irritability of the labellum, in several distantly-allied forms, is highly remarkable. When touched it is described as quickly moving. This is the case with some of the species of Bolbophyllum."[2] We have not, however, considered it well to multiply examples, since to understand and appreciate them some special knowledge of the structure and anatomy of orchid flowers is essential. As our remarks on the spontaneous movements of plants are drawing to a close, we may allude to the subject of a memoir by Pro-

---

[1] "Gardener's Chronicle," 1841, p. 348.
[2] Darwin, " Fertilization of Orchids," p. 172.

fessor Caspary, on the effect of extreme cold in producing movements in the branches of trees in frosty weather. The amount of motion seems to be directly proportionate to the intensity of the cold, but how it is produced has not yet been explained.[1]

M. Lecoq has also described certain rhythmical tremors in the leaves of *Colocasia esculenta*, to which Dr. Masters has directed attention. These are stated to occur at intervals, the plant in the meantime being perfectly at rest ; so violent are the vibrations that on one occasion the pot in which the plant was growing shook so violently that it could with difficulty be steadied. This statement has also been confirmed by another observer. The emission of water from a pore near the apex of the leaf has been occasionally observed in this plant, and it has been suggested that the tremors may have been occasioned by the efforts of the plant to rid itself of the water. But, as Dr. Masters remarks, it is certain that in many cases no such aperture is visible in the plant in question, and that the emission of water is not by any means a common phenomenon.[2]

---

[1] Caspary in "Report of Proceedings of Botanical Congress of London in 1866," p. 98.

[2] "Popular Science Review," vii., p. 25. *Limnocharis Humboldtii* is said to have also a terminal pore at the apex of each leaf from which superabundant moisture drains off.

CHAPTER XI.

SLEEP OF PLANTS.

As long ago as the time of Pliny, nearly two thousand years, certain appearances in plants were observed, which seemed to indicate a condition of repose, as exhibited in the phenomena since designated as the "sleep of plants." The celebrated Linnæus devoted an essay to this subject, and it has since been made the theme of various authors. By the "sleep of plants," is generally included such movements of the leaves as take place periodically towards the close of the day, and which consist in moving upwards or downwards into such a position that the blade of the leaf shall be vertical or nearly so. There is no real analogy between the sleep of animals and the sleep of plants, and the latter term must therefore be accepted rather as a poetic simile, than as a record of fact. The term "nyctitropism" has been proposed as a less objectionable synonym, but its use will scarcely be necessary here, since the more popular term is not liable to be misapplied.

"The fact that the leaves of many plants place themselves at night in widely-different positions from what they hold during the day, but with the one point in common, that their upper surfaces avoid facing the zenith, often with the additional fact that they come into close contact with opposite leaves or leaflets, clearly indicates, as it seems to us, that the object gained is the protection of the upper surfaces from being chilled at night by radiation. There is nothing improbable in the upper surface needing protection more than the lower, as the two differ in function and structure." [1]

We are prepared to accept this as the most feasible reason for the night movement of leaves, strengthened as it is by experiments which were made in this direction, and which demonstrated that leaves which had been fixed in a horizontal position during nights when the temperature was below freezing-point, suffered much more from injury by frost than other leaves on the same plant, which were permitted to assume their usual vertical position.

As it is an important feature to determine what is the probable reason for this "sleep movement" in leaves, we give verbatim the results of one experiment. "We exposed on two occasions during the summer, to a clear sky, several pinned-open leaflets of *Trifolium pratense* which naturally rise at night, and of *Oxalis purpurea* which naturally sink at night (the

---

[1] Darwin, "The Movements of Plants," p. 284.

plants growing out of doors) and looked at them early on several successive mornings after they had assumed their diurnal positions. The difference in the amount of dew on the pinned-open leaflets, and on those which had gone to sleep was generally conspicuous; the latter being sometimes absolutely dry, whilst the leaflets which had been horizontal were coated with large beads of dew. This shows how much cooler the leaflets fully exposed must have become, than those which stood almost vertically, either upwards or downwards, during the night." [1]

When a seed germinates in the ground, the first pair of leaf-like organs which appear above the surface are usually different in form and texture from the true leaves which are afterwards developed.

These primary organs are the cotyledons, and between them the delicate little bud, or plumule, of the succeeding leaves nestles. This remark applies generally to the large number of plants which have a pair of cotyledons. Up to the period when Dr. Darwin commenced his experiments, from the record of which we have already quoted, sleep movements had only been observed in the cotyledons of two plants. This observer has determined, however, that these movements take place in a large number of plants. In some instances the cotyledons sleep,

---

[1] Darwin, " Movements of Plants," p. 293.

whilst the leaves do not. In others the cotyledons do not move, whilst the leaves are remarkable for their sleep movements. In some plants the cotyledons rise vertically upwards, whilst the leaves move downwards, and in others the reverse takes place. Reflecting on the reason why this movement should take place in the cotyledons, this author is of opinion that the movement, by means of which the blade is made to rise or fall almost vertically at night, has been acquired for some special purpose, and he does not doubt that purpose to be the protection of the upper surface of the blade, and perhaps also of the little central bud, or plumule, from radiation at night.

No sane person, reading over the details of experiments like these, would hesitate in the conclusion that, whatever the reason may be, there must be some adequate and sufficient cause for the movements which the cotyledons and other plant organs exhibit. That such movements are merely matters of chance is a position wholly untenable, but their orderly and systematic recurrence leads at once to the conclusion that, however much we may differ in our speculations as to the probable reason why such movements take place, there will be no doubt in our minds that they are the agents in some work which is beneficial to the plant which exhibits them. The theory suggested seems to be a sufficient one, and not opposed to known facts, and it appears to be one which is justified by the evidence. Under these circumstances we see no difficulty in accepting it as an interpretation.

We return now to the sleep of leaves. This is a phenomenon not unknown to those who have been in the habit of observing plants, as even children have noticed it. The different appearance which certain leaves present in the evening to that of mid-day could not escape recognition. Such, for instance, as the drooping leaflets of the common acacia-tree, the Robinia of gardeners, and the reflexed leaflets of the wood-sorrel, and white clover. These are the most common and readily observed. Investigation has shown that movements of leaves, upwards or downwards, in the evening, are far more common than has been supposed; to some of the details of which we shall have to refer. Movements of this kind must not be confounded with the closing of flowers on dull days, or at eventide, nor with their turning towards, or in opposition to the sun. For the present we concern ourselves entirely with the periodic leaf movements which are observed night and morning, and which bear an intimate relationship to the elevation and depression of the cotyledons, to which we have already directed attention. The movements called by Darwin the " nyctitropism of leaves."

The pretty little wood-sorrel (*Oxalis acetosella*) is almost universally known and admired. The leaves are of a beautiful pale green, not unlike in form the common white clover. Each leaf consists of three lobes or leaflets, which are heart-shaped, attached at the base at the top of a slender erect footstalk. During the day these leaflets are spread out nearly flat, radiating equally from the top of the footstalk. In the evening each leaflet gradually falls, until the under side nearly touches the footstalk, and, in so doing, the leaflets which are broadest upwards, are strongly bent inwards, so that each side is deeply concave. In this condition the leaves remain throughout the night (fig. 38). Leaves of this plant were carefully watched for their periodic movements. After half-past five in the evening the leaflets sank rapidly, and at seven o'clock depended vertically, and remained nearly the same until the morning, when by a quarter to seven they had commenced to rise, and continued rising for an hour. Between eleven o'clock in the morning and half-past five in the afternoon they moved four times up and down before the last great fall for the night commenced. The rising and falling during the day was slight as compared with the nocturnal fall. The highest point was reached at

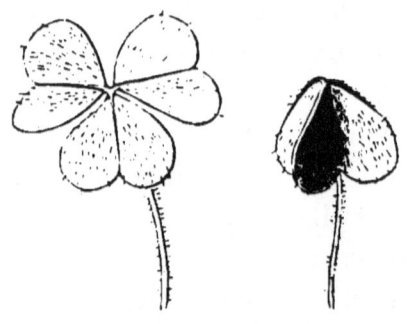

Fig. 38.—Leaves of wood-sorrel.

noon. This is one of the best plants to obtain for persons who reside in town, and desire to watch the movements of the leaves. It will grow freely and readily in a small flower-pot, by the window, in a sitting room, covering the top with its delicate green leaves, so as to be a pretty object, as well as a most interesting one. Several other species are commonly cultivated, and many of these exhibit similar phenomena, although some of them do not.

The blimbing (*Averrhoa bilimbi*) is an oriental fruit tree, which may be mentioned here because of its belonging to the same natural order as the humble little wood-sorrel. This tree and some of its movements were known a hundred years ago.[1] The leaves move spontaneously during the day, they move also in response to a touch, being what is termed "sensitive," and finally they subside into a condition of sleep at night. It is said to be a remarkable sight to observe the leaflets of this tree sinking rapidly one after the other, and then slowly rising. At night the leaflets hang down vertically, and are then motionless. By regulating the light in a conservatory, the behaviour of a plant under variations of light was observed. A leaflet was seen to rise in diffused light for twenty-five minutes, and then a blind was removed so that a strong light fell upon it, and within a minute the

---

Dr. Bruce in " Philosophical Transactions for 1785," p. 356.

leaflet began to fall. The descent was accomplished by six descending steps, that is, by falls succeeded by a slight rising, so as to cause a kind of oscillation, each fall being greater than the rise. The plant was again shaded, and a long slow rise commenced, which continued until the sunlight was again admitted. It is unnecessary to enter into all the minute details of this experiment, the object being to show that a rise and fall of the leaflets took place as a consequence of the increase or diminution of direct sunlight.[1] In another chapter we have alluded to the "sensitive" movements of another exotic fruit-tree belonging to the same genus.

The leaves of the common clover (*Trifolium repens*) are similar in size and form to those of the wood-sorrel, but of a darker green. Their movements nevertheless are quite dissimilar, for the leaflets instead of descending at night, rise and fold over each other so that the under surface is exposed. As evening approaches the two side leaflets twist themselves and move towards each other until their upper surfaces come into contact. At the same time they bend downwards until their midribs form an angle of about forty-five degrees with the upper part of the footstalk. The terminal or middle leaflet rises without twisting, and bends over until it rests upon and forms a kind of

---

[1] Darwin, "Movements of Plants," p. 332.

roof over the edges of the closed side leaflets. By this means a kind of cone is formed with the apex towards the footstalk (fig. 39). The middle leaflet in this movement passes through an angle of 90° to 140°. The falling of the middle leaflet in this "clover" was observed on the morning of two days. On the first day the leaflet fell between eight o'clock in the morning and three in the afternoon. On the second day it fell between seven o'clock in the morning and one in the afternoon. After this fall the leaflet began to rise again, but slowly, until four o'clock, when the rapid rise for the evening commenced.[1] It is interesting to compare these two little plants with trifoliate leaves, and observe how the same end is attained, by two diverse and opposite means, the one by rising, the other by falling; but both equally decided and remarkable.

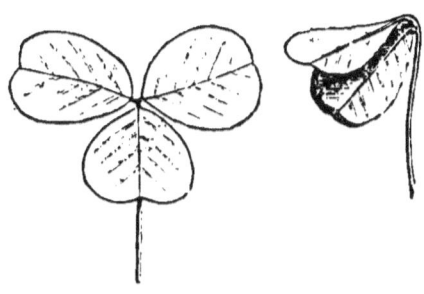

Fig. 39.—Leaflets of clover awake and asleep.

The melilot is another plant with trifoliate leaves, and this exhibits a different class of movements in its nocturnal gyrations. Darwin has thus described the process which he observed. "The three leaflets

---

[1] Darwin, "Movements of Plants," p. 352.

of each leaf twist through an angle of ninety degrees, so that their blades stand vertically at night, with one lateral edge presented to the zenith. We shall best understand the other and more complicated movements if we imagine ourselves always to hold the leaf with the tip of the terminal leaflet pointed to the north. The leaflets in becoming vertical at night could, of course, twist so that their upper surfaces should face on either side ; but the two lateral leaflets always twist so that this surface tends to face the north, but as they move at the same time towards the terminal leaflet, the upper surface of the one faces about N.N.W., and that of the other N.N.E. The terminal leaflet behaves differently, for it twists to either side, the upper surface facing sometimes east, and sometimes west, but rather more commonly west than east. The terminal leaflet also moves in another and more remarkable manner, for whilst its blade is twisting and becoming vertical, the whole leaflet bends to one side, and invariably to the side towards which the upper surface is directed : so that if this surface faces the west, the whole leaflet bends to the west, until it comes into contact with the upper and vertical surface of the western lateral leaflet. Thus the upper surface of the terminal and of one of the two lateral leaflets is well protected."[1] This com-

---

[1] Darwin, "Movements of Plants," p. 345.

plicated movement, which it is rather difficult to comprehend at once from a description, is another example of the prodigality of means by which in nature the same end is accomplished.

The common chickweed (*Stellaria media*) has the leaves in pairs, opposite to each other, and as night approaches these leaves rise upwards with their faces towards each other, the uppermost pair but one closing over the terminal pair, and thus the growing point is protected.

The cultivated nasturtium (*Tropæolum*) has a very simple nocturnal motion, which is not observed unless the plants have been well illuminated during the day. These leaves have naturally a tendency to turn to the sun, so that when growing in full light the blade of the leaf is sloped. At night, however, they become vertical by the bending of the footstalk at about an inch below the blade of the leaf. In the morning they resume again their diurnal position. Were it not the fact that these leaves maintain a vertical position during the night, and assume a more horizontal one in the early morning, it might have been thought that the movements in these leaves were only the result of heliotropism, or turning to the sun. A series of experiments has, however, demonstrated that this is not the case, and that the plant has its true nocturnal motion.

Several species of lupins are under cultivation,

and some of these exhibit very peculiar movements. In one class the leaflets bend downwards at night, in another, they rise upwards, and, in a third, partly up and partly down. The last is the most curious, for the leaflets are numerous, and spread out like the fingers of an open hand. There may be eight or nine of these leaflets, those nearest the base being the shortest. At night the shorter leaflets, which face the centre of the plant, are depressed, whilst the longest leaflets on the opposite side are elevated, the intermediate ones being slightly twisted. In this way the leaves form during the night a kind of vertical star, the edges of the leaflets being directed towards the zenith, whereas during the day the position of the leaves is horizontal, with the upper surface turned towards the zenith.

A curious circumstance is related of a plant of the yellow lupin (*Lupinus lutens*), in which different leaves on the same plant went to sleep in a different manner. Two leaves, the leaflets of which at noon stood at about forty-five degrees above the horizon, rose at night to sixty-five or sixty-nine, so that they formed a hollow cone with steep sides. Four leaves on the same plant, which were horizontal at noon, formed vertical stars at night, and three other leaves, which were equally horizontal at noon, had all the leaflets sloping downwards at night.[1] It is difficult

---

[1] Darwin, "Movements of Plants," p. 343.

to propose a satisfactory solution of these phenomena, for in one plant we have an illustration of all the three methods in which different species of lupin have been observed to pass the night.

The French bean (*Phaseolus vulgaris*) also exhibits sleep movements, but only under special conditions. With plants growing out of doors no tendency to sleep was observed in July, whereas in August the same plants had most of their leaflets in a condition of repose. In this plant the leaflets sink vertically at night, whilst the footstalk rises a little. Other species of the same genus have been observed, and all of them sleep in a like manner.

We have given as many illustrations as necessary from ordinary and well-known plants, either native, or in common cultivation, and these we would now supplement by a few observations on foreign plants such as are cultivated in the green-house or conservatory. For the facts we shall be indebted here, as on previous occasions, to Mr. Darwin's recent work, which, like his other works, is a perfect cyclopædia of facts and observations.

In *Desmodium gyrans*, one of the "sensitive plants," so-called, the leaves consist of a large elliptical terminal leaflet, and two very small lateral ones. The large terminal leaflet sinks vertically at night, whilst the footstalk, or petiole, rises. By this rising of the petioles an altered and more compact appearance is given to the plant. The young petioles, near the top of the plant, rise to such an extent as to be nearly parallel to the stem, whilst the older ones rise considerably. In the evening the rising of the petioles is almost completed before the terminal leaflet begins to fall. Whilst the plant is awake—that is, during the day-time, the leaflets are in constant motion; but, after six o'clock in the evening, when the nocturnal descent has commenced, the direction is almost directly and straightly downwards. After the leaflets are completely depressed for the night they move very little or not at all. The little lateral leaflets do not appear to sleep. They were seen in motion, jerking as they usually do, at ten and at eleven o'clock at night. At one o'clock in the morning the leaflets were still jerking rapidly. At half-past three the jerking was not observed. At half-past eight in the morning it had commenced jerking again. "This leaflet, therefore, was moving during the whole night, and the movement was by jerks up to one a.m. (and possibly later), and again at half-past eight a.m."

Similar nocturnal movements, such as the elevation of the petioles, and the vertical depression of the leaves have been observed in several other plants allied to the above, belonging to the same great natural order.

In *Acacia Farnesiana* the difference between the appearance of a waking and sleeping plant is most

remarkable. The leaf is
a very compound one (fig.
40) consisting of a petiole
with about seven pairs of
pinnæ, or secondary peti-
oles, each of which is
feathered with little leaflets. Towards night the
pinnæ move forwards,
and sink downwards.
The leaflets become
directed towards the apex
of the pinna, and over-
lap each other, so that
the "pinnæ then look
like bits of dangling
string" (fig. 41). Mere
verbal description can

Fig. 40.—Leaf of *Acacia Farnesiana* awake.

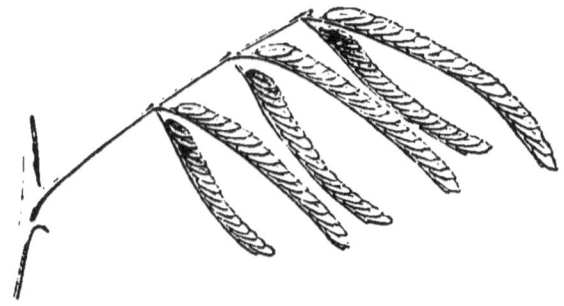

Fig. 41.—Leaf of *Acacia Farnesiana* in a sleeping condition.

give no idea of the contrast between the appearance of this plant as seen by day, and as seen by night.

In *Coronilla rosea* the leaves have nine or ten pairs of opposite leaflets, which during daytime project horizontally. At night these leaflets rise so that the opposite leaflets nearly touch each other, at the same time they bend backwards and towards the stem, sometimes to such an extent that their midribs are parallel to the petiole. This position, both as regards the uprising of the leaflets, and their direction backwards on the petiole, is just the reverse of what usually takes places in the order to which this plant belongs.

In making these experiments it was proved that in order to exhibit fairly their nocturnal movements, the soil of the plants must not be kept too dry. Also that the temperature must not be kept too low; but this would naturally vary in different plants, those which are natives of hot countries requiring a higher temperature to exhibit their natural activity, than those which are denizens of a more temperate clime. A plant out of doors, although in good health, did not exhibit any nocturnal phenomena, whilst a plant of the same species in a warm greenhouse had its leaflets all drooping at night. In the case of many plants it was also found indispensable that the leaves should be well illuminated during the day in order to their sleeping at night. One plant which had its leaves violently agitated by the wind during the day was thereby prevented sleeping by night.

In some instances, as has been detailed, the leaves, or leaflets, are elevated at night, and in others they are depressed. Of the genera examined experimentally there are thirty-seven in which the leaves or leaflets rise and thirty-two in which they sink at night. In a species of *Bauhinia* from Brazil the nocturnal movement was different from any others to which we have alluded. In this plant the leaves are large and broad, with a deep notch at the ends. At night the two halves rise, and close together, with the upper faces closely applied to each other, like closing a book. In young plants the petioles rise also at the same time. Owing to the closing up of the leaves in this manner, the plants have a much more compact appearance at night than during the day.

Without attempting to go over all the argument, either as to the causes which operate in producing these nocturnal movements, or their utility in the economy of the plant, to the latter of which we have already alluded, we may refer to one point which has not been evident from our narrative, but which is insisted upon by Mr. Darwin in his summary. This is in reference to the continuous movement, to a greater or a less extent, during night as well as day, except where the close pressure or imbrication of the leaves prevents motion.

"Any one who had never observed continuously a sleeping plant," he says, "would naturally suppose that the leaves moved only in the evening when going to sleep, and in the morning when awaking; but he would be quite mistaken, for we have found no exception to the rule that leaves which sleep continue to move during the whole twenty-four hours; they move, however, more quickly when going to sleep, and when awaking, than at other times. That they are not stationary during the day has been demonstrated. It is troublesome to observe the movements of leaves in the middle of the night, but this was done in a few cases; and tracings were made during the early part of the night of the movements in the case of several plants,[1] and the leaves after they had gone to sleep were found to be in constant movement. When, however, opposite leaflets come into close contact with one another, or with the stem at night, they are, as we believe, mechanically prevented from moving, but this point was not sufficiently investigated."[2]

It is very certain that the more, and the closer, the growth of plants is investigated, the more evident does it become that there is a continual motion of some kind going on, and that a state of life is a state

---

[1] In *Oxalis*, *Amphicarpæa*, two species of *Erythrina*, *Cassia*, *Passiflora*, *Euphorbia*, and *Marsilea*.

[2] Darwin, "Movements of Plants," p. 403.

of motion. We have only of late years began to appreciate the fact that plant organs are capable of spontaneous movement, and although the earlier chapters in the history of such phenomena have been written, these are, doubtless, only preliminary to fuller and more elaborate details which the future will reveal.

The number of genera in which nocturnal movements have been observed is not more than about eighty-six. All the species in a genus will, as a rule, exhibit the same kind of motion. Here and there an erratic species may be found expressing its movements in the language of another genus, but usually they are tolerably uniform. The large natural order of *Leguminosæ* or pod-bearers, to which the pea and bean belongs, contains the largest number of genera possessed of distinct nocturnal movements. It is in this order that the "sensitive plants" are located. There are in fact more genera belonging to this order in which "sleeping partners" have been observed than in all the other families put together.

It is difficult to establish a definite boundary between plants which exhibit nocturnal movements and those which exhibit similar movements, but in a less pronounced degree, as for instance in those which elevate or depress their leaves but slightly, or those which only exhibit the same kind of rotation in the evening and morning as at mid-day. This

S

feature adds force to the suggestion made by Darwin, that the movements which are termed "nocturnal" or "sleep" movements, are caused in the same manner, and only differ in degree from the ordinary rotation of leaves and branches. In fact "nyctitropism" or the sleep of leaves is "merely a modification of their ordinary circumnutating movement, regulated in its period and amplitude, by the alterations of light and darkness."

Viewed in any aspect, and under any name, these phenomena are most interesting and curious. Whether we choose to call them "sleep," or only "periodic," it matters not, the facts themselves cannot but enlarge our ideas of plant existence. If we have been in the habit of considering that in its manifestations of activity the vegetable world is much the inferior of the animal world, investigations of this kind will compel us to modify our somewhat rash generalisations, and look with increased respect on the organisms we have unduly depreciated.

> Floral apostles, that, in dewy splendour,
> Weep without woe, and blush without a crime,
> Oh, may I deeply learn, and ne'er surrender
> Your lore sublime.

CHAPTER XII.

METEORIC FLOWERS.

IT was a happy idea of Linnæus to construct a "floral clock," with the hours representing the opening or closing of certain flowers. It was also the same botanist who applied the name of "meteoric flowers" to such as closed and expanded periodically, at, or near the same period of time, or such as appeared to be influenced especially by definite atmospheric changes in opening or closing. Pretty and poetical as such a theory may be, it is doubtful if it extends beyond this. So many circumstances may modify the periodicity to such an extent as to upset any horological arrangement. A dull day and a bright sunny one, a dry morning or a moist one, will certainly not produce the same results. The opening and closing depending so much on light and temperature will be related more to the bright, clear sky, and the warm genial atmosphere, than to the particular hour of the day. Admitting all these influences and conditions, it is doubtless true that under a normal condition there are many flowers which open or close nearly at the same time, or within an hour. It might be said that certain flowers have a manifest tendency to open or close at, or about a certain time, unless this tendency is disturbed or thwarted by special interference. Probably this was all that Linnæus ever intended, and that his design was to indicate that some flowers expanded with the first break of day, others not until noon, and others again in the evening, or during the night. Thus much is undoubtedly true and in this aspect the subject has its interest, sufficient to demand a brief notice at our hands, although practically of but little scientific value.

The catalogue, as constructed by Linnæus[1] would be too long for insertion here, and we shall be content with enumerating the abbreviated list as corroborated by De Candolle,[2] the hours of waking being first stated.

2 a.m.	Purple Convolvulus (*Ipomœa purpurea*)
3–4 a.m.	Fior de Notte (*Ipomœa nil*)
„	Great Bindweed (*Calystegia sepium*)
4–5 a.m.	Goatsbeard (*Tragopogon pratense*) and other Cichoraceæ
5 a.m.	Yellow Arctic Poppy (*Papaver nudicaule*)

---

[1] Linnæus, "Philosophia Botanica," 2nd ed., Vienna, 1783, p. 273.
[2] De Candolle, "Physiologie Végétale," ii., 484.

METEORIC FLOWERS.

5–6 a.m.	Nipplewort (*Lapsana communis*)
,,	Common Blue Convolvulus (*Convolvulus tricolor*)
6 a.m.	Spotted Cat's Ear (*Hypocharis maculata*)
6–7 a.m.	Sow-thistle (*Sonchus*) and Hawkweed (*Hieracium*)
7 a.m.	Water Lilies (*Nymphæa* and *Nuphar*)
7–8 a.m.	Venus's Looking-glass (*Specularia speculum*)
8 a.m.	Scarlet Pimpernel (*Anagallis arvensis*)
8–9 a.m.	Nolana (*Nolana prostrata*)
9 a.m.	Marigold (*Calendula arvensis*)
9–10 a.m.	Red Sandwort (*Arenaria rubra*)
10–11 a.m.	Fig Marygold (*Mesembryanthemum nodiflorum*)
11 a.m.	Lady Eleven o'clock (*Ornithogalum umbellatum*)
12 a.m.	Blue Passion Flower (*Passiflora cœrulea*)
2 p.m.	(*Pyrethrum Corymbosum*)
5–6 p.m.	Night Flowering Catchfly (*Silene noctiflora*)
6 p.m.	Evening Primrose (*Œnothera biennis*)
6–7 p.m.	Marvel of Peru (*Mirabilis jalapa*)
7 p.m.	White Evening Lychnis (*Lychnis vespertina*)
7–8 p.m.	Night-flowering Cereus (*Cereus grandiflorus*) and some others.

Although light and temperature appear to have an influence on the opening and closing of flowers, yet in some species at least, which open or close near a definite time, there is a decided effort made to open or close when the appointed hour arrives, even when the external conditions are unfavourable. De Candolle made some experiments with flowers excluded from the daylight, and found that in some cases a plant which was accustomed to flower in daylight at a certain time, still followed the clock hour in their opening and closing.

Fritsch made some extensive observations at Prague on the opening and closing of flowers.[1] He was of opinion that there is no hour of the day when the flowers of some plant do not open, and in the majority of instances they are closed at sunset. With the exception of a few hours near midnight, there is no hour at which blossoms do not begin to close—the maximum about six o'clock in the evening. In a few plants the complete expansion lasts only an hour, sometimes not so much. In those blossoms which are fully expanded in the morning the duration of expansion is short. In those blossoms which expand in the afternoon, the condition of waking is limited by the length of time the sun is

---

[1] See "Journal of Horticultural Society of London," vol. viii., p. 1.

above the horizon. In those blossoms which are fully expanded in the night the duration of sleep is shortest. Flowers which are fully expanded in the morning open more rapidly than they close. Those which open in the afternoon close more rapidly than they open. For other facts relative to light, tempera-

Fig. 42.—Scarlet Pimpernel (*Anagallis arvensis*).

ture, and other influences on the movements of flowers, we must refer to the memoir from which we have quoted.

The pretty *Arenaria rubra*, says Edwin Lees, that opens its purple petals wide before the mid-day sun closes them instantly as soon as plucked, or folds them close should a storm obscure the welkin with dark clouds. The daisy "goes to bed," as it is said, before the sun goes down, but the bright yellow wort (*Chlora perfoliata*) closes its flowers before five in the afternoon, and the yellow goat's beard (*Tragopogon pratensis*), so common now in upland meadows, even before noon—hence its colloquial name "Go-to-bed-at-noon." The little pimpernel (*Anagallis arvensis*) sullenly keeps its scarlet petals closely shut on a cloudy or a rainy day, and this so constantly and certainly, that it has been called the "Shepherd's weather-glass," for whatever the barometer may indicate, if the red pimpernel has its flowers expanded fully in the morning, there will, to a certainty, be no rain of any consequence on that day, and the umbrella and the macintosh may be dispensed with.

> Come, tell me, thou coy little flower,
>   Converging thy petals again,
> Who gave thee the magical power
>   Of shutting thy cup on the rain,
> While many a beautiful bow'r
>   Is drenched in nectareous dew,
> Seal'd up is your scarlet-tinged flower,
>   And the rain peals in vain upon you?
>
> The cowslip and primrose can sip
>   The pure mountain dew as it flows,
> But you, ere it touches your lip,
>   Coyly raise your red petals and close;

> The rose and the sweet briar drink
> With pleasure the stores of the sky,
> And why should your modesty shrink
> From a drop in that little pink eye?[1]

Many of the night-blooming flowers are fragrant; it would seem that because they expand at a period when, on account of the darkness, beauty of appearance would be futile to attract, they compensate for this by diffusing delicious odours. The great water-lily (*Victoria regia*) is really nocturnal, as are some others of the water lilies, and the magnificent night-blooming cereus (*Cereus grandiflorus*) is one of the finest, as well as one of the most fragrant of flowers.

The evening primrose (*Œnothera biennis*) is very unlike a primrose, except in colour. Here and there a blossom may be seen expanded in the daytime, but the majority of the flowers do not open till six or seven o'clock in the evening, and then they are slightly fragrant (fig. 43).

Equally well known are the varieties of the "marvel of Peru" (*Mirabilis jalapa*), sometimes called the "four o'clock," although the time for opening is as late as the evening primrose. The "lady of the night" (*Mirabilis dichotoma*) is probably the original "four o'clock," as it opens earlier, but popular names must not always be applied with

---

[1] "Botanical Looker-out," p. 168.

rigid exactitude. These are also pleasantly scented.

Nothing can be more beautiful amongst our native

Fig. 43.—Evening Primrose (*Œnothera biennis*).

wild flowers than a field in which the night-blooming catchfly (*Silene noctiflora*) grows in abundance.

METEORIC FLOWERS. 267

During the day no trace of the plant can be seen, but at seven o'clock in the evening there is a remarkable change. As though called up at the stroke of a fairy wand, the little blossoms, sparkling like gems, are scattered thickly over the ground. Such a sight is not readily forgotten. Let the experience of lovers of flowers expand this theme upon our brief introduction.

There is another aspect of flowers, for which no special provision is made in our arrangement of chapters, but which is of equal interest, and that is the extraordinary form which the floral organs assume in many families of plants. We might denominate such instances as are present to our mind, as the "eccentricities of flowers." The link which unites them to "meteoric flowers" is very slender, but they are closely related to the object of this volume.

The "reason why" such strange forms are assumed cannot always be determined, but in many instances there is an undoubted connection between the form and the object to be attained, namely the fertilisation of the flower. For instance, it has been suggested that the bright patch of colour on the petals of the rhododendron is so placed with reference to the inclination of the stamens, that insects, attracted by the bright colour, and flying directly towards it, must come into contact with the anthers, and disperse the pollen. Dr. Darwin has indicated in his books, and especially in that on the fertilisation of the orchids, how various modifications in the form and structure of flowers are related to the functions they have to perform. He considers that all the so-called "eccentricities" of form in the floral envelopes are useful, or necessary, to the plant. Some very striking illustrations might be adduced in support of this view, which it would be difficult to controvert. It is much more reasonable to assume that the strange forms sometimes met with are designed to overcome some difficulty, or attain some definite purpose, than to regard them as mere vagaries, or freaks of nature, indulged in by chance, or out of mere sportiveness.

There are no plants which can compete successfully with the orchids for singularity of form. They have been termed the "monkeys of the vegetable world," and not a very happy designation, for monkeys do not assume fantastic forms, although they may perform fantastic tricks. Some of these plants grow on the ground in the usual manner, others are parasitic on living trees, the most luxuriant and singular are natives of tropical countries. The lower petal, or labellum in particular, is liable to endless variation. In some cases it is slipper-like in form, in others trumpet shaped. Here and there the entire flower resembles an insect, sometimes to such a degree as to have originated a popular name, as suggestive of the resemblance. This is the case in our own bee orchis (*Ophrys apifera*), not uncommon on chalky downs (fig. 44). In one curious species the appearance is that of a bird on the wing. The "snipe orchis" is just such a flower as in semi-barbarous countries or amongst a superstitious people, would be associated with some mystic legend, as in the case of another orchid, the "Holy Ghost" plant, The horned labellum of a New Guinea orchid (*Pachystoma*) exhibits another type or irregularity (fig. 46). And the figure of a flower of a species of *Dendrobium*, a parasitic orchid from New Guinea (fig. 47) will represent a curious form in which two of the strap-shaped, erect petals, resemble the horns of an antelope. The zebra orchis (*Oncidium zebrinum*) from Venezuela, has white

Fig. 44.—Bee Orchis (*Ophrys apifera*).

Fig. 45.—Snipe Orchis ("Gardener's Chronicle").

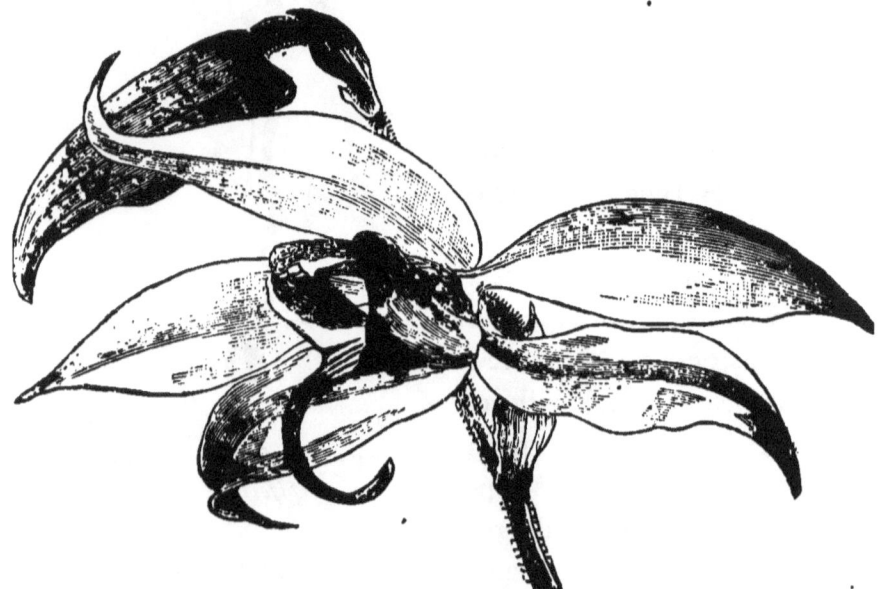

Fig. 46.—Flowers of *Pachystoma Thomsoni* ("Gardener's Chronicle").

flowers, with violet transverse bars across the petals, and may be characterised rather as singularly beautiful than eccentric.

Of all genera of orchids none are stranger than that called *Masdevallia*. The species are designated by such names as "chimera," nycterina, troglodytes, &c., which are justified by their weird and "uncanny" appearance. Description alone could scarce convey an adequate impression of these strange flowers, some

Fig. 47. — *Dendrobium D'Albertisii* ("Gardener's Chronicle").

with long tails hanging down from the extremity of each petal, others with similar appendages thrust out, more rigidly, in all directions, and all with a sombre hue, suggesting thoughts of gorgons, medusæ, and of "hydras and chimeras dire."

The flowers of the family of birthworts are tubular or trumpet-shaped, here and there strangely inflated, lurid in colouring, sometimes fœtid in odour, and often of large size; many of them are climbers, and their rugged, contorted stems, with a snake-like form have in many countries a reputation as an antidote to snake-bites.

The asclepiads have regular flowers, often with

thick fleshy petals, sometimes resembling wax flowers, and a structure interesting to botanists because of its

Fig. 48.—Zebra Orchis (*Oncidium zebrinum*).—" Gardener's Chronicle."

departure from the ordinary type, the pollen masses

being of a similar nature to those of Orchids. Our illustration of a *Ceropegia* (page 192) exhibits the vase-like shape which those flowers assume, whilst others are much more simple, and scarcely conspicuous.

The "hand plant" of Mexico (*Cheirostemon platanoides*) has acquired its designation from the stamens being extended like the five fingers of a hand, from a large calyx, like a leather cup, true petals being absent. The flowers secrete a quantity of liquid like sugar and water, tasting and smelling like toast and water. Each flower continues about a fortnight in perfection before it begins to fade. It was narrated of this flower, when first found in 1787, that it was so great an object of curiosity with all the inhabitants of New Spain, that the flowers were gathered with avidity by the Indians, even before their full expansion, and thus the seeds were not allowed to ripen. The tree was venerated from time immemorial by the Indians, who believed it to be a solitary tree, of which no other existed or could exist in the world. Nevertheless other trees were discovered in Guatemala in 1801.[1]

Side-saddle flowers (*Sarracenia*) are surmounted by a kind of hood, not unlike a parasol, with the petals hanging out, all round the margin, like little saddle-

---

[1] "Botanical Magazine," plate 5,135.

flaps. Of a different character, but no less strange, are the laterally flattened pink flowers of a plant now common in gardens, which first bore the name of "Dutchman's breeches" (*Dielytra spectabilis*). Some of the tubular flowers are beautiful enough to merit the old belief that they were the habitations of the "good people."

> 'Twas I that led you thro' the painted meads,
> Where the light fairies danced upon the flowers,
> Ranging on every leaf an orient pearl,
> Which, struck together with the silken wind
> Of their loose mantles, made a silver chime.

NOTE.—By an unfortunate accident the manuscript of this, and the five or six succeeding chapters, was lost on its way to the printers, and had to be re-written under disadvantages, for the notes and memoranda accumulated during some fifteen years had been incorporated, and the originals destroyed. Undoubtedly some omissions will have to be accounted for by this circumstance.

CHAPTER XIII

HYGROSCOPISM.

HYGROMETRIC and Hygroscopic are two terms which have been applied indiscriminately, or interchangeably, to indicate certain movements in the parts of plants resulting from a susceptibility to dryness or moisture. These phenomena are often exhibited by dead and dried organs, but sometimes during vitality. It is difficult to mark a distinct line between such phenomena as exhibited by dead and living tissue, nor is this essential, since in all cases the causes are similar, and consist in the different size, form, and density of subjacent series of cells, which expand and contract, at different rates, and to diverse extent, by absorption or loss of moisture, thus producing twisting, curving, or contortion in alternate directions. In other words, it may be accounted for "by supposing that the cells on one side are larger, and have thinner walls than those on the other; and these will therefore be most easily distended when placed in water, and will soonest lose their fluid in drying."

One of the oldest and best known illustrations of hygroscopism, is the awn of the wild oat (*Avena fatua*), which, in times gone by, has been taken advantage of by designing men to impose on the credulous and superstitious. These awns are twisted in their lower portion, and so susceptible of moisture, even that of the human breath, or a damp hand, that they at once exhibit spontaneous movement, twisting and writhing as if endued with animal life. "Jugglers in the good old time predicted events, and told fortunes, from its motions; and, to cover the cheat, they called the awn 'the leg of an Arabian spider,' or 'the leg of an enchanted fly.'" The true rendering of the phenomena, when it came to be understood, supplanted the jugglers. Hooke, one of the early writers on microscopical objects, saw beneath the mystery, for he writes:—

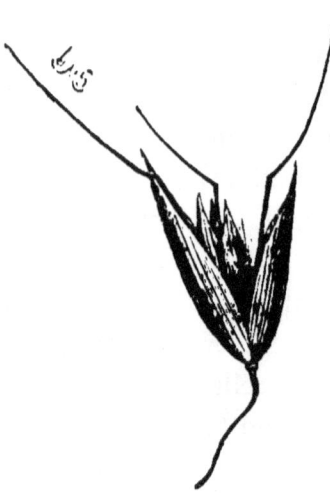

Fig. 49.—Wild oat (*Avena fatua*).

"Its sensibility to changes in the atmosphere seems to depend on the different texture of its parts, for the awn is composed of two kinds of substances, one that is very porous, loose, and spongy, into which the watery streams of the air may be very easily forced, which will be thereby swelled and extended in its dimensions; and a second that is more hard and close, into which the water can very little or not at all penetrate, this therefore retaining always very near the same dimensions, and the other stretching and shrinking, according as there is more or less moisture or water in its pores, by reason of the make and shape of the parts the whole body must necessarily unwreath and wreath itself."[1] Another grass, which, although not a native, is often cultivated, has very long awns, which are subject to twisting and writhing under increase or decrease of moisture. The whole structure and mode of action in the awns of this species were made the subject of an elaborate investigation by Mr. Francis Darwin.[2] The seed terminates downwards in a sharp, strong, oblique point, armed with a dense plume of barb-like hairs; upwards it is continued in a strong, woody awn, of which the lower part is strongly twisted on its own axis, and its upper portion untwisted and fringed with a series of beautiful hairs, so as to impart a feathery appearance. It is bent like a knee between the twisted and untwisted portions. When the seed is fixed, and the awn free, moisture applied to it causes the lower portion to untwist, and with it the

---

[1] Hooke, "Micrographia," p. 151.
[2] "On the Hygroscopic Mechanism by which Certain Seeds Bury Themselves," "Linnæan Transactions," 2nd series, vol. i., p. 149.

feathered upper part is carried round, so that the movement is conspicuous. As the moisture evaporates, the twisting of the lower portion of the awn again takes place, and the twisting and untwisting may be repeated at will, as moisture is applied or withheld. If the feathered end of the awn is fixed, and the seed is free, the latter will be carried round, rotating with the movement of the twisting or untwisting awn. The object of the investigations alluded to was to determine what was the reason for this twisting, and what purpose it served in the economy of the plant. Without entering into the details, which may be consulted at will, it may be assumed as proved that the hygrometric property possessed by the awn, whereby it twisted and untwisted, would enable the sharp point at the lower extremity of the seed to penetrate and bury itself in the ground. It was shown by experiment "that the seed was buried, both as it untwists, and also as it returns to a state of torsion. By a combination of these two processes the awn is thrust into the soil to such a depth as to cover up the seed completely." A seed entangled in the branches of a low bush, and left out of doors for eight days, had buried itself to a depth of thirty-one millimètres, or nearly double the length of the seed, impaling a piece of rotten leaf in its way. It was found that seeds dropped from a height of a few feet usually preserved a nearly vertical position, striking the ground with the point. If allowed to fall among low vegetation they become fixed in a more or less oblique position, the seed resting on the ground. The length of the feather renders entanglement easy, and, when a seed is once entangled, the hairs serve to hold it fast and prevent the wind blowing it away.

The movement in the awn of *Stipa spartea*, an allied species in the Red River colony, cause the sharp, rigid points of the seed to enter and bury themselves in the wool of the sheep with which the grass comes into contact.[1] Further than this, it is affirmed that the seeds penetrate the skin by their screw-like movement, and cause the death of the animals. At the same time that hygroscopism receives an illustration by these grasses, its utility is demonstrated by the dissemination of the seeds, and continuance of the species. In the Geranium family, after the ovules are fertilised, the centre of the receptacle continues to grow until it is prolonged into a long beak, with the seeds arranged around the base, and the elongated styles applied to the sides of the beak. The peculiar beak-like form which the fruit thus assumes has acquired for the plants the popular names of Crane's-bill, Stork's-bill, &c. The mode in which the carpels are loosened at the base, and curl upwards like a watch-spring to the top of the beak, is

---

[1] See Museum No. 2, Royal Gardens, Kew.

familiar to all who have observed the members of the family. That the twisting is the result of an hygrometric property in the awn has been demonstrated by Professor Asa Gray.[1] He says: "The narrow carpel is pointed at the base; the long awn or style in drying bends at right angles with the carpel, and twists in many turns, depending on the amount of dryness, and untwists in a moister air or when wet. We had wondered that no one seemed to have given an account of the way in which this mechanism acts so as to bury the seed in the ground. Dispersed by the wind over the loose or sandy soil which these species prefer, the seed bearing end, being the heavier, lies next to the ground, and is the comparatively fixed point around which the long awn makes circular sweeps, whether in twisting or untwisting. This gives a rotary movement to the carpel, fixes the sharp end in the soil, and, whether twisting or untwisting, causes it to bore into and bury itself in the ground."

M. Roux says that in *Erodium*, when the seeds are thus interred, the moisture of the soil soon destroys the epidermis, and this allows the long beak to detach itself at its articulation with the style, leaving it planted in good condition quietly to germinate. Thus then, it may be seen that, by their own hygro-

---

[1] "Use of Hygrometric Twisting of the Tail to the Carpels of *Erodium*," "American Journal," 3rd series, vol. xi., 1879, p. 153.

scopy, the seeds become their own planters, and effectually secure themselves in a favourable position to ensure the continuance of the species.

The little cruciferous plant to which the name of "rose of Jericho" has been applied (*Anastatica hierochuntina*), has a divided claim to be included with hygroscopic, and also with mystic plants. It is a native of the dry wastes of Northern Africa and Palestine, and the sandy deserts of Arabia. It is a small bushy plant, not more than five or six inches high. After flowering the leaves fall off, and the branches become dry, shrivel, and curve inwards towards the centre, so as to form the plant into a kind of ball. In this condition it is easily uprooted from the soil, carried by the winds, blown and tossed across the desert into the sea. Upon coming into contact with water the plant again unfolds itself, the branches expand, the seed vessels open and disperse the seeds, which are carried by the tide and deposited on the shore. The property of expanding when in contact with moisture led to a superstitious regard for the plant which, it was believed, expanded on the anniversary of the birth of our Saviour. It was called also *Rosa Mariæ*. The plant may be kept for years, if preserved in a dry place, but at any time when the root is placed in water, or the entire plant immersed it will expand and, it is said, in the course of a few hours the buds of the flowers will swell, and appear as if newly taken from the ground. Old Gerard calls it the "heath rose of Jericho,"[1] but he says that "the coiner spoiled the name in the mint, for of all plants that have bin written of, there is not any more unlike unto the rose." He gives a description similar to that recorded above, with figures of the plant in the dried and also in the expanded state.

Akin in its movements to the *Hierochuntina* is the "wind witch" of the Russian steppes, so graphically described by Schleiden: "In autumn the stem of the thistle plant rots off and the globe of branches dries up into a ball, light as a feather, which is then driven through the air by the autumnal winds, over the steppe. Numbers of such balls often fly at once over the plain with such rapidity that no horseman can catch them, now hopping with short, quick springs along the ground, onward in a spirit-like dance over the turf, now caught by an eddy, rising suddenly a hundred feet into the air. Often one "wind witch" hooks on to another, twenty more join company, and the whole gigantic, yet airy mass rolls away before the piping east wind."[2]

The designation "rose of Jericho" has been applied to the capsular fruits of a species of fig marygold, from the Cape of Good Hope (*Mesembry-*

---

[1] Gerard, "Herbal," lib. 3, p. 1,386.
[2] Schleiden, "The Plant," p. 354.

anthemum tripolium). On the approach of rain, or when placed in water, these seed vessels gradually open like a star, closing again as they become dry. Thunberg, remarking on this circumstance, says:— "In this we see the wisdom of an All-wise Creator, inasmuch as this plant, which is found in the most arid plains of South Africa, keeps its seeds closely

Fig. 50.—Capsules of *Mesembry-anthemum tripolium* closed.

Fig. 51.—Capsule of *Mesembryanthemum tripolium* open.

locked up in time of drought; but when the rainy season comes, and the seeds can grow, it opens its capsules, and lets fall the seeds, that they may be dispersed abroad." These capsules have also been known by the name of "Flowers of Crete."

The violent dehiscence of fruits is occasioned in

many instances by the hygroscopism of some of the parts. Although this usually takes place in dead tissues, some of the instances are of interest in this connection. One of the best known is the fruit of the sand box-tree (*Hura crepitans*). The capsule is about the size of an orange, and consists of a number of carpels, packed together side by side. When dry the carpels separate, and fly apart with a loud report. When once separated they cannot be compressed again into their original form. These capsules were formerly used as "sand-boxes," before the invention of blotting paper, but had to be bound together in order to prevent their sudden dehiscence. It was not unusual for them to fly in pieces after many years.

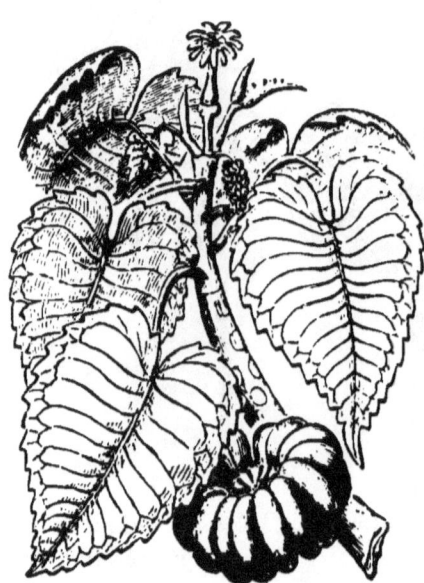

Fig. 52.—Sand-box (*Hura crepitans*).

The pods of some plants of the pea and bean family (*Leguminosæ*) have a tendency to separate at the valves, and become twisted or curl backwards with considerable force. The large pods of an

African tree (*Pentaclethra macrophylla*) possess this property in an exaggerated degree. When fastened together by strong wires they break themselves in pieces in their efforts to become free.[1] The length of these pods is from twenty-two to twenty-five inches, with a breadth of about three and a-half inches. From the observations of Professor Oliver, it has been ascertained that the increase and decrease of length between dryness and moisture is sixteen per cent., so that the contraction in one pod would not be less than three inches. Contractility of a similar character but to a less extent has been observed in a plant of the cucumber family (*Echinocystis lobata*).[2]

An illustration of a different kind to those hitherto adduced is furnished by Dr. Darwin, in one of his works. It refers to a plant which bears a name in allusion to its hygrometric predilections (*Porlieria hygrometrica*). "In the Botanic Gardens at Wurzburg," he says, "there was a plant in a pot, out of doors, which was daily watered, and another in the open ground which was never watered. After some hot and dry weather there was a great difference in the state of the leaflets on these two plants; those on the unwatered plant, in the open ground, remaining half, or even quite, closed during the day.' But twigs

---

[1] Oliver, in "Linnæan Transactions," xxvi., p. 415.
[2] Wyman, "Proceedings American Academy," iii., p. 167.

cut from this bush, with their ends standing in water, or wholly immersed in it, or kept in damp air, under a bell glass, opened their leaves though exposed to a blazing sun, whilst those on the plant in the ground remained closed. The leaves on this same plant after some heavy rain, remained open for two days; they then became half closed during two days, and after an additional day were quite closed. This plant was now copiously watered, and on the following morning the leaflets were fully expanded. The other plant, growing in a pot, after having been exposed to heavy rain, was placed before a window, with its leaflets open, and they remained so during the daytime for forty-eight hours, but after an additional day were half closed. The plant was then watered, and the leaflets on the two following days remained open. On the third day they were again half closed, but on being again watered remained open during the two next days. From these facts we may conclude that the plant soon feels the want of water, and that as soon as this occurs, it partially or quite closes its leaflets, which in their then imbricated condition expose a small surface to evaporation. It is probable that this sleep-like movement, which occurs only when the ground is dry, is an adaptation against the loss of moisture."[1]

---

[1] "The Movements of Plants," p. 337.

The hygroscopic character of some of the species of *Selaginella* is familiar to horticulturists. These plants have somewhat the appearance of large mosses, and are not uncommon in greenhouses. In the classification there is one entire section devoted to the species which have the foliage curved inwards when dry, so that many of them roll up by contraction into the form of a ball. This is the habit of *Selaginella convaluta*, a species abundant in Bahia and Pernambuco, and also of *Selaginella lepidophylla*, which latter has been called the "Resurrection plant," from its habit of expanding under moisture. One of the earliest observers of this phenomenon in South America was the celebrated traveller Martius, who called the plant *Lycopodium hygrometricum*.

Sensibility to variations in humidity is also an important factor in the dispersion of the spores in many of the ferns. In these the sporangia are girt by an elastic ring, which assists in the rupture of the sporangium. "When the sporangia arrive at maturity, and are under certain favourable conditions as to dryness, the elasticity of the ring causes them to burst open with force and sound sufficient to be heard, and this takes place in a direction at, or very near to, a right angle with the direction of the ring."[1] This serves to remind us that the bursting of the spathe

---

[1] Smith, "Ferns, British and Foreign," p. 51.

in some palms, with a loud report, as sometimes recorded, will be due to a like cause. The dryness of the atmosphere inducing contraction in one series of cells, greater than in another, produces a violent rupture, as in the separation of the carpels of the "sand-box," accompanied by a sharp sound.

Microscopists hold in great favour an object which finds a place in almost every "cabinet," and consists of the peristome of the common hygrometric moss (*Funaria hygrometrica*). In this, and many other species, especially *Hypnum*, the urn-shaped receptacles which contain the spores are at first covered with a lid, or hood, and when the latter falls away are seen to be fringed with a single or double row of teeth, called the peristome. These teeth converging inwards cover the spores, and prevent their escape, when expanded or recurved they permit of the free discharge of the contents of the urn. This fringe is exceedingly sensitive to moisture, opening and closing when breathed upon, or as the moisture of the breath evaporates. It is a very pretty and available illustration of a vegetable hygrometer. Some bryologists object to this as a legitimate inference. They assert that the movement is not vital, but is merely mechanical, resulting from the diverse character of the outer and inner layer of cells, of which the peristome is composed. Admitting the structure to be thus accurately described, it becomes a structural adaptation to secure a certain end, which is beneficial to the plant. Surely it must be too delicate a distinction to admit specialised structure in other instances, such as stigmatic surfaces, &c., and reject it in this. As the opening and closing of the peristome takes place whilst the plant is living, even whilst the urn of the plant is still attached and living; and as it is of a manifest utility in securing the dispersion of the spores at such a period as when the moisture of the atmosphere would best secure their germination, we are prepared to retain the peristome of mosses as a satisfactory illustration of hygroscopism.

In the Liverworts (*Hepaticæ*) the spores are mixed in the capsules with spiral threads, or elaters. If the contents of one of these capsules are moistened after they have become dry, the spiral threads will be seen wriggling and twisting about, by means of the relaxation of the spiral, such movement also being of assistance in the dispersal of the spores. In the same manner we have observed the threads in such Myxomycetes as *Trichia*, in which the threads are spiral, relax a little when moistened after they have become dry, but, in this instance, only to a limited extent.

The examples we have given are sufficient to show that, inasmuch as there are movements in plants which result from the influences of light, temperature, and other causes, as demonstrated in preceding chapters; so also there are movements which are dependent upon the slightest variation in the humidity of the surrounding atmosphere. The further we investigate the phenomena of plant-life, the more do we become assured that, if there is not "a soul in every leaf," there is at least a marvellous adaptation of the parts, like a well-ordered machine, in order to secure definite and essential results.

> In all places, then, and in all seasons,
>   Flowers expand their light and soul-like wings,
> Teaching us, by most persuasive reasons,
>   How akin they are to human things.
>
> And with child-like, credulous affection,
>   We behold their tender buds expand:
> Emblems of our own great resurrection;
>   Emblems of the bright and better land.

CHAPTER XIV.

DISPERSION.

IN many cases there is undoubtedly a close relationship between the form which fruits or seeds assume, and the mode by which they are dispersed. If it be true, as some contend, that the ultimate object of every plant is its own perpetuation, then the dispersion of the seed is an important operation, which consummates all other acts, and it would be anticipated that adequate provision would be made to ensure its full attainment. This is certainly not accomplished by any uniform method, but through various agencies, and in a multitude of ways. We shall be able in some cases to comprehend distinctly how the operation is performed, whilst in others it is more complex, and sometimes obscure. One agency in dispersion is the wind, which wafts seeds that are provided with wings to their destination. Another agency is undoubtedly water in which they are floated to a congenial spot. Another is the application of local force, by elasticity or hygroscopy, by means of which the seeds are forcibly expelled. Another is specialised structure by aid of which various animals are utilised as a means of transport. In these, and various collateral ways, we are enabled to associate the modifications of form with the mode of distribution, and it is to a few of the most striking examples of these different modes that this chapter is devoted. Illustrations will also be found scattered through other portions of this volume incidentally, when given in association with other subjects, and especially in the chapter on "mimicry," in which the same type of structure, and, presumably, the same mode of dispersion will be found repeated in different and widely separated orders. Neither here nor there have we assumed the exhaustion of so fertile a theme.

When writing of hygroscopism we have already alluded to the facility with which the seeds are dispersed from such fruits as burst open with violence and jerk out their contents, as in the sand-box tree. In a similar manner we might have instanced *Byttneria aspera*, one of the order *Sterculiaceæ*, but a more important tree in many respects is the Mahogany tree (*Swietenia mahogani*), the woody fruits of which separate so freely, and disperse the seeds, that it is difficult to meet with any but fragments of the capsule in the countries where it flourishes. Another advantage is possessed by this tree, in that the seeds themselves are winged, like a samara, so that two modes of dispersion are combined in the same fruit.

The balsams, which a few years ago were great favourites in country districts, and ornamented every cottager's window, scatter their seeds to a great

Fig. 53.—Balsam (*Impatiens*).

distance by the violent rupture of the fruits, in allusion to which circumstance they have been called "Touch-me-not."

A remarkable instance of violent dehiscence is to be seen in a member of the cucumber family, named from its habits the "squirting cucumber." (*Momordica elaterium*). The small oval fruit, not more than an inch and a half in length, is covered with prickles. When quite ripe this separates from the stalk, and jerks out its juice, with which the seeds are mixed, from the little opening at the base. The fruit was formerly considered valuable as a medicine, and was cultivated for that purpose in this country, but now that its use has been superseded it is rarely to be met with.

The witch hazel of North America (*Hamamelis virginica*) exhibits a peculiar elasticity in the seeds, or embryo of the seeds, which are thrown out with such force as to strike people violently in the face who pass through the woods. Collecting a number of the capsules, and laying them on the floor, Mr. Meehan found that the seeds, or embryos were thrown out, generally to the distance of four or six feet, and in one instance as much as twelve feet.

We might here descend a little lower in the vegetable world than we have done hitherto, and indicate amongst the lower cryptogamia one or two very decided instances of the forcible ejection of the mature fruit corpuscles. In fungi, for example, we have observed the grass all round a spot in which a number of cups of a "bird's nest fungus" (*Crucibu-*

lum vulgare) were growing, sprinkled to the height of six or eight inches with the ejected sporangia; and this is a species in which the sporangia are attached to the cup by an elastic cord, so that forcible ejection is not a recognised means by which distribution is affected. Yet no other explanation can be offered for the occurrence of the sporangia on the surrounding grass. Another species (*Sphærobolus stellatus*) normally ruptures at the apex, and expels the globose sporangium, like a cannon ball, but no larger than a small pill into the air. Still more minute, and in another section, the little mucor-like fungus, which grows so profusely on cowdung, ejects the little sporangia to an enormous distance, in proportion to the size of the plant. We have seen them covering the leaves of vines, and other plants, in minute specks, like the dung of flies, at an almost incredible distance from their source. Although at first received with some doubt, C. B. Plowright has affirmed that the spores of some agarics do not simply fall from the hymenium, but are ejected, in some manner not yet explained, for three or four inches, not only in a line with, but above the plane from whence they proceed. We have since been able to corroborate the fact, in two or three instances, but without succeeding in tracing the cause, or being able to submit a reasonable theory to account for the phenomenon. Inasmuch as we were at first sceptical, with others, it is with greater pleasure that we recant.

Passing now to atmospheric dispersion, the seeds of composite plants are transported from place to place, chiefly by means of the pappus with which they are crowned. Although varying in different species, this coronet has but one purpose, the dispersion of the plant. "Elevated on the apex of a long beak, the parachute of the seed of the goatsbeard (*Tragopogon pratensis*) consists of a number of slender spokes, which diffuse themselves circularly, and are "telarly interwoven," somewhat after the fashion of the spider's web. This comparatively intricate structure is given as a countervail to the great size and weight of the seed. The down of the dandelion is supported on a long and slender pedicle, and is an object of vulgar admiration; but it scarcely equals in beauty the similarly patterned fruit of the helminthia. The thistle's-down is, on the contrary, sessile—the threads being sometimes only spinous, at other times plumed like a feather—and the down of the latter is peculiarly light. The coronet of the carline thistle is remarkable for its elegance and circular spread and plumage, and buoys easily its silky-coated seed. In the sow thistles what we most admire is the ribbed and striated seeds, but the down that diffuses them is abundant and of pure whiteness. The seeds of the coltsfoot afford an example of a structure, common in the order, where the seed is surmounted by a tuft of silken hairs armed, at regular intervals, with a series of denticles or spines, only visible with a good magnifier. We have a contrast to this in the curious fruit of the blue-bottle (*Centaurea cyanea*) which has a small tuft of asbestine spines at the base, and a large but short tuft of rigid stout lanceolate spines on the top, the edges of each of them indented with close and sharp serratures like a saw. This tuft cannot float the seed in the air, but it will obviously direct and hasten its descent into the soil, and it will be remarked that the forward direction of the spines must be opposed to every influence to cast them up again, after having been buried under the surface."[1]

The stalks of the down in the dandelion contract closely together in moist and wet weather—a beautiful provision to secure its dispersion only in a dry day, when it is driven off by every zephyr, and not unoften by the schoolboy, who thus endeavours to resolve his doubts as to the hour :—

> Dandelion, with globe of down,
> The school-boy's clock in every town,
> Which the truant puffs amain
> To conjure lost hours back again.

The dispersion of seeds by means of a coronet of delicate hairs is not confined to composite plants.

---

[1] Johnston's "Botany of the Eastern Borders," p. 126.

A like provision may be observed in the willow herbs (*Epilobium*), and in many of the *Apocynaceæ*, as well as the *Asclepiadaceæ*. Yet this is only one provision for the dispersal of seeds by the agency of the wind. Another, and equally successful contrivance is the expansion of the sides of the seed into a membranous wing. These winged seeds reach their highest development in the trumpet flowers (*Bignoniaceæ*), where the large wings extend three or four inches, and the seeds float like a large butterfly, wafted from place to place, until a secure home is reached. In our own country such winged seeds are usually minute, if we exclude the heavier and less delicate winged fruits of such trees as the maple, ash, and elm, which are called *samaræ*. These latter are doubtless aided in their dispersion by means of their wing-like margins, to which we shall have occasion to refer again hereafter, when writing of the similarities which prevail in the organs of very diverse plants. We are justified, then, in asserting that special provision is made for the dispersion of many seeds through the air, by means of the wind.

Spiny fruits are found amongst the members of many of the families in the vegetable kingdom. It is evident that the rigid spines with which they are armed aid considerably in their dispersion. The name of "caltrops" has been applied to some of these, in allusion to the "calcitrapa" which was employed in ancient warfare to harass the enemy's cavalry. One of these kind of caltrops (*Tribulus terrestris*) is widely diffused, probably on account of the facility with which the fruits are transported

Fig. 54.—Caltrops, or fruits of *Tribulus terrestris*.

in the wool of animals. They have an elegant, symmetrical, star-like form, and the spines are very sharp and rigid. Another, but less complex, fruit (*Pedalium murex*) has its dispersion provided for in a similar manner. The name of caltrops has also been

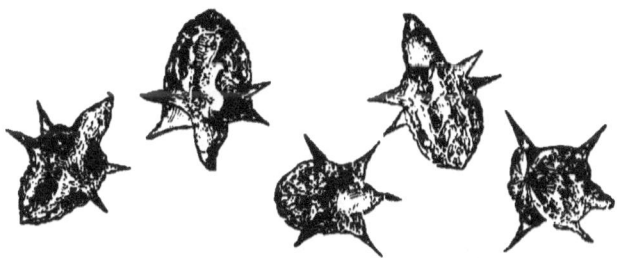

Fig. 55.—Fruits of *Pedalium murex*.

given to one of our indigenous plants, called also "starthistle," on account of the sharp spines of its woody involucre (*Centaurea calcitrapa*). More efficient still are the recurved hooks with which some of the spines of fruits are terminated. To a limited extent this may be seen in the small fruits of the carrot, and a few other of the Umbelliferæ, but much more strongly developed in the "burrs" of the burdock

Fig. 56.—Burdock (*Lappa minor*)

(*Lappa*). We are familiar enough with the tenacity with which these "burrs" will adhere to the clothing of any one passing amongst the plants, but their entanglement in the woolly coat of animals is much more complete. Similar burrs are produced by *Xanthium strumarium*, which is not a native plant, but has been introduced by the pertinacious adhesion of its fruits to the coats of animals. An allied common tropical species (*Xanthium spinosum*) has by a similar means spread itself over a wide area, becoming thereby a nuisance in some of its adopted homes. In South Africa it has established itself, by means of the merino sheep, and "extended itself through the sheep-walks of the colony to such a degree, and so endangered the character of the wool through its achenes, that special legislative enactments have been made in regard to its extirpation, and rigid enforcement of penalties alone has kept it from being a sweeping curse to the wool-producers. In the Orange River Republic, where only until last year (1872) this weed was allowed to revel undisturbed, save where some stray Dutch boer was given less to coffee-drinking and sleep, and more to an intelligent regard for the future of his pasturage, it had so affected the wool of some parts of the country as to make it nearly unremunerative as a staple product. Tardy legislation on the obnoxious introduction had to be adopted there also."[1]

---

[1] Dr. John Shaw in "Journal of Linnæan Society," xvi., p. 202.

The "cleavers," or "goose-grass" (*Galium aparine*), found in every hedge and thicket has small fruits which are densely covered with minute hooks, and transportation is rendered easy. It is in warmer climates, where hooked fruits attain a larger dimension, that they present a formidable appearance. In one of these (*Martynia diandra*) the pair of hooks are very sharp and rigid, the points entering the flesh like a needle ; but even these are exceeded by another species (*Proboscidea Jussieui*),

Fig. 57.—Hooked fruits of *Martynia diandra*.

which the late Frank Buckland was wont to declare must have been created for the express purpose of sticking to the tails of the wild horses that roam the plains of South America. The horns in this species are often five or six inches in length, and the aspect may be readily imagined from our reduced figure (fig. 58). The same family contains the Grapnel plants, of which one species (*Harpagophytum leptocarpum*) is found in Madagascar, and another in Africa (*Harpagophytum procumbens*). The latter, and most effective of the two, although least formidable in appearance, has the capsule armed with a number of rigid woody thorns, standing out in all directions, their tips furnished with two or three recurved hooks like miniature grapnels. It is easy to imagine how such a fruit may be transported, the difficulty being rather to believe in the possibility of its remaining at rest (fig. 59).

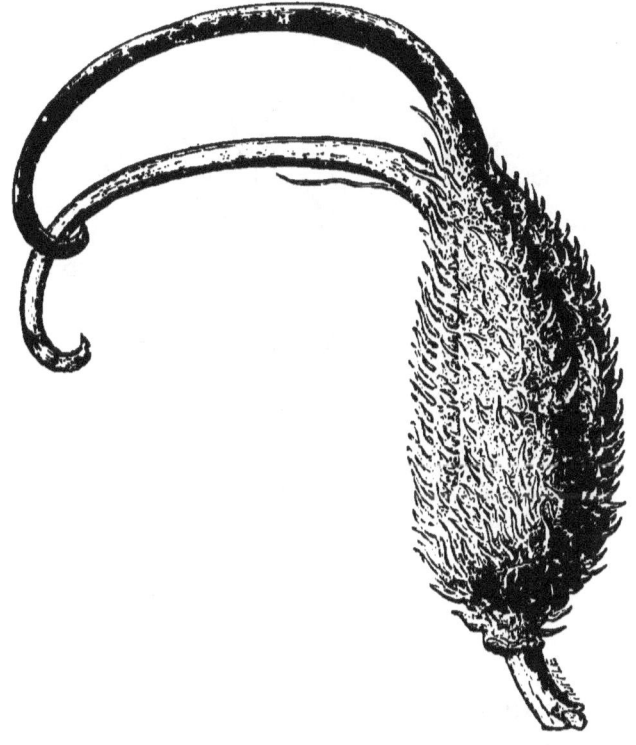

Fig. 58.—Fruit of *Proboscidea Jussieui* reduced.

There could scarcely be a more conclusive evidence of the utility of a large national collection of such objects as those which are the subject of this chapter than a visit to the museums at Kew Gardens would afford. Instead of a hasty glance at mere curiosities, we would suppose the visitor to foster some such design as to go in search of specialised forms of fruits which would be serviceable in dispersion ; or to trace,

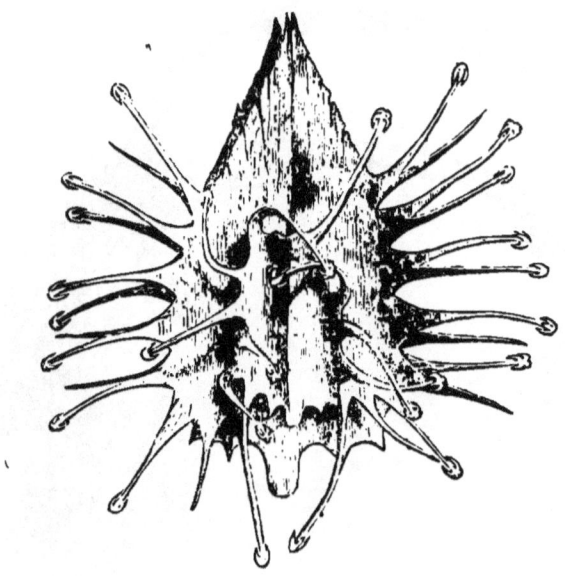

Fig. 59.—Fruit of Grapnel plant, natural size (*Harpagophytum leptocarpum*).

in passing from case to case and from order to order, the recurrence of similar forms in diverse families, so like as to suggest mimicry. With such an object, only to be gratified by such an institution, we would venture to affirm that not only would a visit afford far greater satisfaction, and would excite more

DISPERSION.

intense pleasure, but would also considerably increase the visitor's own appreciation of the educational value of an exhibition too often looked upon, we fear, without so much as leaving a trace on the memory.

What interpretation is to be assigned to the peculiar form of the fruits in the different species of *Trapa*, or "water-chestnut," is not so clear. The plants float in the water, and the fruit of one species (*Trapa*

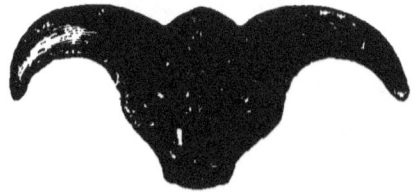

Fig. 60.—Fruit of *Trapa bicornis.*

bicornis), commonly cultivated in China, resembles the head and horns of a bull. Another species (*Trapa bispinosa*) is largely cultivated in Cashmere. In this

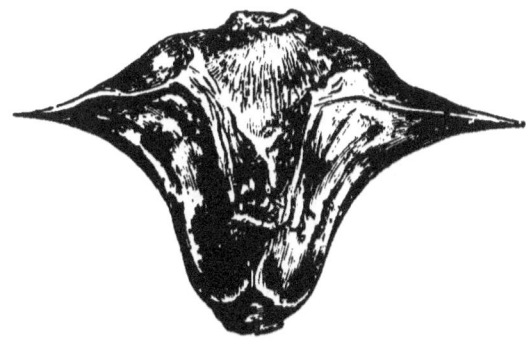

Fig. 61.—Fruit of *Trapa bispinosa.*

fruit the bull's horns are replaced by acute spines. The latter is cultivated to such an extent that it is said to constitute a large portion of the food of the inhabitants, and yields about £12,000 a year in revenue. Moorcroft asserts that from 96,000 to 128,000 ass loads are yielded by the lake of Ooller.[1] There is still another species, with rather smaller fruits, in which four long rigid spines are placed nearly in the same plane, at right angles to each other (*Trapa quadrispinosa*). Had they been produced on land instead of water, it might fairly have been assumed that these appendages would have assisted in dispersion, but under the existing circumstances their utility is not so evident.

The suspension or retention of seeds in a favourable position, until in a fit condition to germinate, has been observed in some cases in a manner so marked as to suggest a special contrivance for the perpetuation of the species. A provision of this nature has been recorded in a plant of the sedge family, native of New Zealand (*Gahnia xanthophylla*), which consists in the filaments of the stamens, which are at first short, and afterwards greatly lengthen themselves. When the ovary is ripened so as to form the nut containing the seed, it is detached from the investing scales, and would fall to the ground if

---

[1] Royle's " Illustrations of Himalayan Botany," p. 211.

it were not caught and retained by the long filaments. It is probable that the object is to obtain a more perfect maturity of the seeds before they drop to the ground.[1]

There is probably some relation in principle between the suspension of the seeds in *gahnia* to what takes place amongst the members of one or two families of trees. For instance, the mangroves, which inhabit the swamps on the margins of great rivers, generally retain their seeds suspended to the branch until after germination has commenced, and when they drop it is into the soft mud, where they immediately take root. In like manner some of the Magnolias have the seeds suspended at the end of the umbilical cords from the margins of the carpels, presumably in order that they may reach a proper degree of maturity before they fall. Examples of this kind are not numerous, but

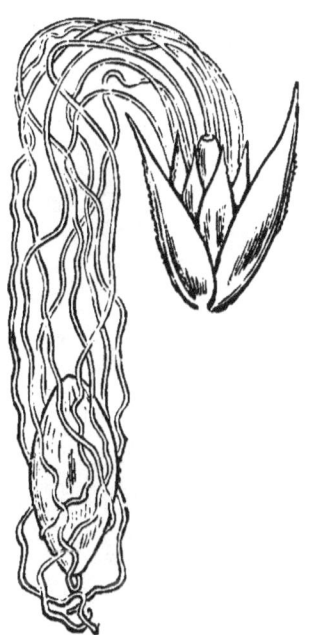

Fig. 62.—Fruit of *Gahnia xanthophylla* suspended. "Gardener's Chronicle."

---

[1] "Gardener's Chronicle," December 13, 1873.

sufficient to afford some explanation of the phenomenon first alluded to. Sir James Smith says, in allusion to the Egyptian bean (*Nelumbium speciosum*), " in process of time the receptacle separates from the stalk, and, laden with ripe oval nuts, floats down the water. The nuts vegetating, it becomes a cornucopœia of young sprouting plants, which at length break loose from their confinement and take root in the mud."

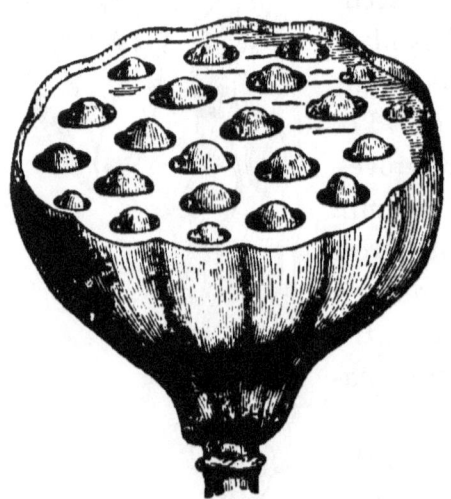

Fig. 63.—Receptacle of the Egyptian Bean (*Nelumbium speciosum*).

The most remarkable of tropical fruits, in their structural aspect, are some of the myrtle family, the seeds of which are enclosed in a large woody urn, or capsule, like a drinking-vessel with a movable lid. In some of them the fruit is no larger than a small walnut, in others as large as a man's head.[1] In some the form is elegant and urn-like, in others it resembles

---

[1] See descriptions and figures of a large number of species in "Transactions of the Linnæan Society," vol. xxx., p. 157, &c.

DISPERSION. 309

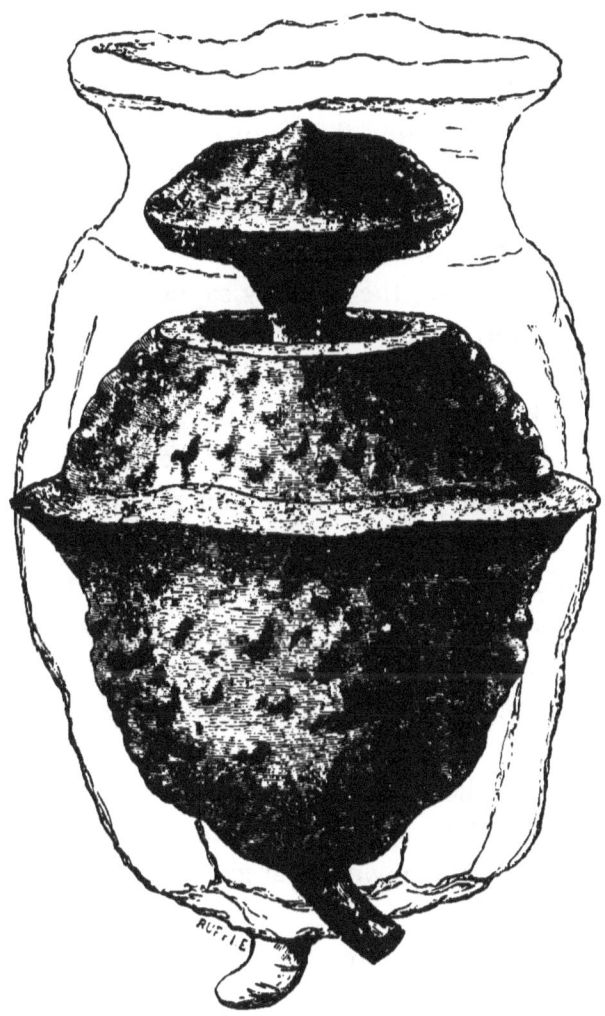

Fig. 64.—Monkey pots (*Lecythis sp.*).

a vase (fig. 64). They are produced on large forest-trees, and are common throughout South America, but especially in the forests of Brazil. The monkeys are exceedingly fond of the delicious "sapucaia" nuts which are produced within these capsules. As Kingsley writes:—"The great urn-shaped fruits, big enough to serve for drinking-vessels, each kindly provided with a round wooden cover, which becomes loose, and lets out the savoury sapucaia nuts inside, to the comfort of all our 'poor relations.' Ah, when will there arise a tropic Landseer to draw for us some of the strange fashions of the strange birds and beasts of these lands?—to draw, for instance, the cunning, selfish, greedy grin of delight on the face of some burly, hairy, goitred old red howler, as he lifts off a 'monkey-cacao' cover, and looks defiance out of the corners of his winking eyes at his wives and children, cousins and grandchildren, who sit round jabbering and screeching, and, monkey fashion, twisting their heads upside down as they put their arms round each other's waists, to peer over each other's shoulders at the great bully, who must feed himself first as his fee for having roared to them for an hour at sunrise on a tree top while they sat on the lower branches and looked up, trembling and delighted, at the sound and fury of the idiot sermon."[1]

---

[1] Kingsley's "At Last," p. 277.

The name of monkey pot, as applied to these fruits, is said to have arisen thus : " When the cup of a Lecythis falls, its lid drops off, the seeds roll out, and it then becomes a hard pot, with a narrow mouth. These pots are used for catching monkeys. Filled with sugar they are placed on the ground which such animals frequent. The sugar attracts the latter, who pick it out leisurely till they are disturbed, when they insert the paw, grasp as much sugar as it will hold, and endeavour to escape with their prize. But their doubled fist being larger than the mouth of the pot cannot be withdrawn, and the monkeys tenaciously holding the sugar, run off with a pot firmly enclosing one paw. This renders it impossible for them to escape from their pursuers by climbing, and they are easily run down."[1] To the credit of the monkeys, it may be added, that it is the young and inexperienced that are caught in this manner, and not the old and wary patriarchs, as intimated by the proverb, common in South America, " He is too old a monkey to be caught with a Cabomba."

The nearest resemblance we have in this country to the structure of the fruits of the *Lecythis* are the comparatively minute and insignificant little capsules of the Pimpernel (*Anagallis*) and the Henbane. In

---

[1] " Gardener's Chronicle," December 28, 1861, p. 1,133.

these instances the seeds are enclosed in a capsule, which opens with a deciduous lid. When the capsule is mature the lid falls off, and the seeds are dispersed. In a remote manner, if such a comparison can be legitimately made, the hood, or cap, of the theca, or capsule, in mosses falls off, and the spores escape, except that they are subject to further retention, until a suitable season, by the incurved teeth of the peristome.

It is easy enough to comprehend the "wherefore" of the monkey pots, and their movable lids, but it is not so evident why, in similar trees of the same family, the seeds should be enclosed in hard, woody capsules, with no orifice, and from which there is no escape, but by the decay of the thick envelope. Such are the Brazil nuts of commerce (*Bertholletia excelsa*). " The fruit, round like a cannon ball, and about the size of a twenty-four pounder, is harder than the hardest wood, and has to be battered to pieces with the back of a hatchet to disclose the nuts. Any one who has hammered at a Bertholletia fruit will be ready to believe the story that the Indians, fond as they are of the nuts, avoid the 'totocke' trees till the fruit has all fallen for fear of fractured skulls."[1] The Capuchin monkeys, according to Humboldt, " are singularly fond of these 'chestnuts of Brazil,' and

---

[1] Kingsley's " At Last " p. 276.

the noise made by the seeds when the fruit is shaken as it falls from the tree excites their appetency to the highest degree."[1] He does not, however, believe the story current on the Orinoco, that the monkeys place themselves in a circle, and by striking the shell with a stone succeed in opening it. That they may try is possible enough, for there is no doubt that monkeys do use stones to crack nuts. The impossibility in this case would be, not in the want of wits, but want of strength; and the monkeys must have too often to wait till the rainy season, when the shell rots of itself, and amuse themselves meanwhile in rolling the fruit about, vainly longing to get their paws in through the one little hole at its base.

Another instance of these wholly closed capsules is the fruit of the cannon-ball tree (*Couroupita guianensis*). This fruit "is a rough brown globe, as big as a thirty-two pound shot, which you must get down with a certain caution, lest that befal you which befel a certain gallant officer on the mainland of America. For, fired with a post-prandial ambition to obtain a cannon ball, he took to himself a long bamboo, and poked at the tree. He succeeded, but not altogether as he had hoped. For the cannon ball, in coming down, avenged itself by dropping exactly on the bridge of his nose, felling him to the

---

[1] Humboldt's "Personal Narrative," vol. v., p. 537.

ground, and giving him such a pair of black eyes that he was not seen on parade for a fortnight."[1]

It has been suggested that, in such trees as the Brazil nut, which is produced in forests swarming with monkeys, that the closed capsule is a protection, and that if the capsule had been an open one, not only would Brazil nuts make less appearance in the markets of the world, but the trees would run a risk of extirpation. We must confess that we are not prepared to accept this as a sufficient reason for the closed capsule. Monkey pots are open capsules, and the trees are not yet extirpated; yet monkeys are as delighted with the "sapucaia" as with the "Brazil nut." In such a case speculation does little good, when it is simply an excuse for one's own ignorance.

Fig. 65.—Cannon Ball (*Couroupita guianensis*).

---

[1] Kingsley's "At Last," p. 275.

DISPERSION. 315

In many cases there appears to be no special provision for the dispersion of seeds, and yet, when duly considered, such a future has not been disregarded, It may be that, covered with a pulpy fruit, attractive to some member of the animal kingdom, the hard seeds have thus been transported to a considerable distance, and found a congenial soil. This fact is recognised by zoologists themselves, as will be evident from the following extract:—" Doubtless many of our most richly-wooded landscapes owe much of their timber to the agency of quadrupeds and birds. Linnets, goldfinches, thrushes, goldcrests, &c., feed on the seeds of elms, firs, and ash, and carry them away to hedge-rows, where, fostered and protected by bush and bramble, they spring up, and become luxuriant trees. Many noble oaks have been planted by the squirrel, who unconsciously yields no inconsiderable boon to the domain he infests. Towards autumn this provident little animal mounts the branches of oak trees, strips off the acorns and buries them in the earth, as a supply of food against the severities of winter. He is most probably not gifted with a memory of sufficient retention to enable him to find every one he secretes, which are thus left in the ground, and springing up the following year, finally grow into magnificent trees. Pheasants devour numbers of acorns in the autumn, some of which having passed through the stomach, probably germinate. The nuthatch in an indirect manner also frequently becomes a planter. Having twisted off their boughs a cluster of beech-nuts, this curious bird resorts to some favourite tree, whose bole is uneven, and endeavours by a series of manoeuvres to peg it into one of the crevices of the bark. During the operation it oftentimes falls to the ground, and is caused to germinate by the moisture of winter. Many small beeches are found growing near the haunts of the nuthatch, which have evidently been planted in the manner described."[1]

Not only do the birds and small quadrupeds assist in the dispersion of seeds in the way just indicated, but even to a much greater extent. As, for instance, we have seen amongst the droppings of birds the small undigested seeds of pulpy fruits which they devoured, which seeds retained all their powers of germination, especially of elderberries and mulberries. This mode of dispersion is undoubtedly a very extensive one in practice. Neither can we ignore altogether the service which some insects may render in the dispersion of minute organisms. To what extent this may be carried it is difficult to determine, but we may give an illustration. It is not an unusual circumstance to find, when examining a species of black mould or of *Torula*, growing on rotten wood, that

---

[1] "The Zoologist," p. 442.

insects have been at work, destroyed the threads, and left behind them characteristic cylindrical exuviæ. These undoubtedly had passed through their bodies, for fragments of threads were mixed with the spores. On two or three occasions we have determined that such spores still possessed the power of germination, even perhaps in an increased degree. These insects may assist in the dissemination of such spores, as molluscs do in feasting on the gills of an agaric, and then retiring to the shelter of some prostrate trunk. These speculations, however, concern a very minute class of organisms which, as a rule, we have deemed it prudent to ignore in this volume.

Then there are larger animals which contribute their share to the dissemination of plants, and especially those of the human family. It would be impossible to enter fully on such a topic, at the end of a chapter, but one or two brief suggestions may be permitted. Even to the present day, writes Schleiden,[1] are marked the footsteps of the bands of nations which in the middle ages emerged from Asia into Central Europe, by the advance of the Asiatic steppe plants, such as the kochia and the Tartar sea-kale, the former into Bohemia and Carmola, the latter into Hungary and Moravia. The

---

[1] Schleiden, "The Plant," p. 301.

North American savage significantly calls our plantain (*Plantago major*), or road weed, the "footstep of the whites," and a common species of vetch (*Vicia cracca*) still marks the former abode of the Norwegian colonists in Greenland. One of the most striking instances of this kind is the gradual extension of the thorn apple over the whole of Europe, which has followed the bands of gipsies out of Asia ; this race make frequent use of this poisonous plant in their unlawful proceedings, and hence much cultivated by them, it also occurs, uncalled for, near the place where they have made their habitations. Auguste St. Hilaire says,[1] "In Brazil, as in Europe, certain plants appear to follow in the footsteps of man, and preserve the traces of his presence ; frequently have they helped me to discover the situation of a ruined hut, in the midst of the wastes which extend out beyond Paracuta. Nowhere have the European plants multiplied in such abundance as in the plains between Theresia and Monte Video, and from this city to the Rio Negro. Already have the violet, the borage, some geraniums, the fennel, and others, settled in the vicinity of Sta. Theresia. Everywhere we found our mallows and camomiles, our milk-thistle, but, above all, our artichokes, which, introduced into the plains of the Rio de la

---

[1] Introduction to "Flora of Brazil."

Plata, and the Uruguay, now clothe immeasurable tracts, and render them useless for pasture." After the War of Deliverance, in many places where the Cossacks had encamped, was found the tick-seed, a plant allied to the goosefoots, which is quite exclusively indigenous in the steppes on the Dnieper, and in a similar manner was *Bunias orientale* spread with the Russian hosts, in 1814, through Germany even to Paris.

A curious circumstance has been recorded as this chapter is passing through the press, which deserves permanent record, albeit, it would have been more in place in the seventh chapter. An Indian species of *Loranthus*, which is a parasite like the misletoe, grows on evergreen trees, especially *Memecylon*. The fruit is a viscid pulp, which surrounds the seed and adheres to whatever it falls upon, until the seed germinates. The peculiar locomotion now recorded is confined to the first stage of germination of the seed, and indicates a rambling habit for the purpose of securing a suitable home. "The radicle at first grows out, and when it has grown to about an inch in length, it developes upon its extremity a flattened disc; the radicle then curves about until the disc is applied to any object that is near at hand. If the spot upon which the disc has fastened is suitable, the germination continues, and no locomotion takes place; but if the spot should not be a favourable one, the germinating embryo has the power of changing its position. This is accomplished by the adhesive radicle raising the seed and advancing it to another spot, or, to make the process plainer, the disc at the end of the radicle adheres very tightly to whatever it is applied to; the radicle itself straightens, and tears away the viscid berry from whatever it has adhered to, and raises it in the air. The radicle then again curves, and the berry is carried by it to another spot, where it adheres again. The disc then releases itself, and by the curving about of the radicle is advanced to another spot, where it again fixes itself. This, Dr. Watt says, he has seen repeated several times, so that to a certain extent the young embryo, still within the seed, moves about. It seems to select certain places in preference to others, particularly leaves. The berries on falling are almost certain to alight upon leaves, and although many germinate there, they have been observed to move from the leaves to the stem, and finally fasten there."[1]

---

[1] N. E. Brown in " Gardener's Chronicle," July 9, 1881, p. 42.

CHAPTER XV.

MIMICRY.

IN the animal kingdom certain resemblances between the members of one group and those of another, considerably removed from it in the system of classification, have of late years been the subject of much discussion. On the supposition that these resemblances have been acquired, and are designed to serve some purpose in the economy of nature, the term "mimicry" has been applied to them. Subsequently it has been proposed to substitute another term, that of "homoplasy," but this has not met with general acceptance; we have, therefore, adopted the older term. Mr. H. W. Bates first introduced the subject to notice, with some very striking examples of "mimetic resemblances" in Lepidopterous insects, which have since been much augmented by others. Very few allusions have hitherto been made to such resemblances in plants, although Mr. A. Bennett[1] has opened the question, and with these organisms the subject is still in a very elementary stage. It shall

---

[1] Mimicry in Plants in "Popular Science Review," vol. xi. (1872), p. 1.

be our purpose to indicate such examples as have come to our knowledge, but rather as a record of facts than with any design to theorise about them. In animals it has been contended that the resemblances are acquired by natural selection and the survival of the fittest, such resemblances being for the benefit of the organisms which acquire them. The data are at present insufficient to apply such a theory to plants, but the instances are sometimes so striking and curious that they could not be ignored as remarkable phenomena in plant life.

Of all known plants none are more weird and singular than the Cacti, with their angular succulent stems, armed with spines, and the absence of leaves. In many instances the flowers are large, showy, and beautiful. These plants are numerous in the hotter and drier parts of America. "Sometimes globular, sometimes articulated, sometimes rising in tall polygonal columns, not unlike organ-pipes." Now and then attaining a very gigantic size, occasionally so small that "they get between the toes of dogs." In the similarly dry and arid tracts of Africa these plants are absent, but their place is occupied by species of *Euphorbia*, which resemble in form and habit the Cacti of America. In our woodcut (fig. 66) is represented one of these Euphorbias, growing amongst rocks at the Cape of Good Hope, which is entirely dissimilar from our common wood spurge and the other Euphorbias of temperate regions, whilst it is scarcely to be distinguished from a cactus,

Fig. 66.—Euphorbia, resembling a cactus growing amongst rocks in Damara Land.

except that they have usually small and inconspicuous flowers. The two families of plants are widely separated from each other, almost as far as possible for plants to be, and yet the resemblance is so great that in the absence of flowers it is difficult to believe that they do not belong to one family. Not only do they resemble each other, but they are also "imitated" by plants of another family, the Asclepiadaceæ, of which the species of *Stapelia* might equally be attributed to Cacti or Euphorbia. These plants may be seen growing together in the "succulent house" at Kew.

Before leaving these succulent plants we may also instance certain of the lily family, small aloes of the genus *Haworthia*,[1] in which the fleshy leaves grow close to the ground in the form of a rosette. In this instance the resemblance approaches to that of the house-leek family (*Crassulaceæ*), further removed than even *Euphorbia* and *Cactus*, for one belongs to Monocotyledonous and the other to Dicotyledonous plants.

If from these general features we turn to individual plants, we shall find the number of examples greatly increased. Any one who has had an extended experience will appreciate the difficulties which constantly arise in determining even the "order" of an

---

[1] Two pairs of these plants may be compared—viz., *Haworthia planifolia* with *Echeveria aloides*, and *Haworthia atrovirens* with *Sempervivum arenarium*.

unknown plant in the absence of flowers or fruit. Turn, for instance, again to the *Euphorbiaceæ*, and compare one of the species of *Phyllanthus*, with flattened phyllodes, as *Phyllanthus falcata*,[1] with a similar structure in a species of the Buckwheat family[2] (*Polygonaceæ*). Here, in an unusual form, a striking mimetic resemblance will be encountered.

Or, if we have only the young condition, without flowers or fruit, of such a floating plant as *Jussiæa repens*, one of the Onagraceæ, we shall at once be struck with its resemblance to a similar condition of an Euphorbiaceous plant (*Phyllanthus fluitans*), and, at the same time, with such a cryptogam as *Salvinia rotundifolia*. In our figures of these three plants the resemblance is less striking than in the plants themselves (fig. 67). All of them float on the water, under similar conditions, in different parts of the world.

Dr. Berthold Seemann speaks of having seen, in the Sandwich Islands, a variety of Solanum (*S. Nelsoni*) which looked for all the world like a well-known Buettneraceous plant of New Holland (*Thomasia solanacea*), "the resemblance between the two widely-separated plants being quite as striking as that pointed out in Bates's "Naturalist on the

---

[1] "Botanical Register," pl. 373.
[2] *Muhlenbeckia platycladium*, "Botanical Magazine," pl. 5,382.

Amazons," between a certain moth and a humming bird.

The circumstance has also been alluded to many times that one of our highest botanical authorities of his day figured a species of *Veronica*, without

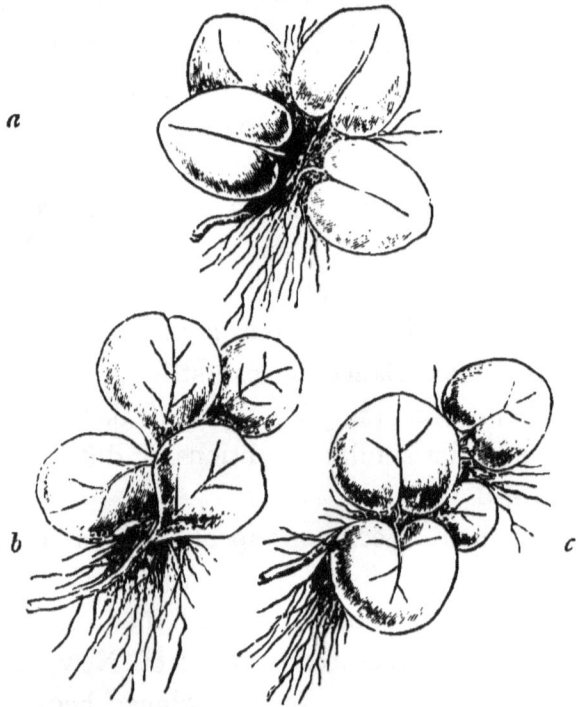

Fig. 67.--Young plants of (*a*) *Salvinia*, (*b*) *Jussiæa repens*, (*c*) *Phyllanthus*.

flowers or fruit, as a conifer. This serves to remind us how many plants there are, scattered through various families, in which the form and arrangement of the foliage is so much like that of a cypress, that it is almost impossible to distinguish them from the foliage alone. Others again are nearly identical with some of the larger species of Lycopodiums, cryptogams allied to ferns and mosses. We have figured one example of a compact composite plant (*Azorella*

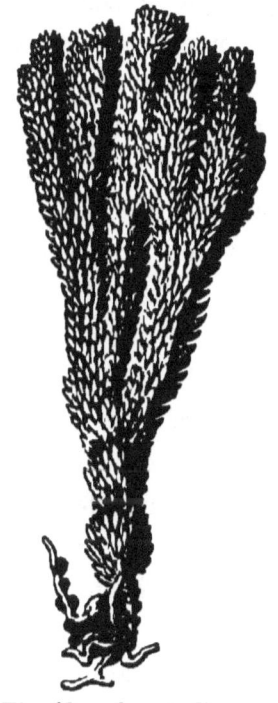

Fig. 68.—*Lycopodium compactum.*

Fig. 69.—*Azorella selago.*

selago) for comparison with one of the club-mosses (*Lycopodium compactum*). Other instances might have been selected from remote families, in which the same resemblance is sustained, so that from the figures it would scarce have been possible to determine whether they were "club-mosses" or not.

Foliage is hardly so satisfactory for comparison, except in cases where the leaves have a strongly marked character, nevertheless we may suggest two or three of the most striking. The leaves of the planes and the maples have a coincidence of type. The pinnate leaves of some of the *Oxalidaceæ* closely resemble, even in their sensitive nature, some of the mimosa family. Or, if we instance individual species, the leaves of the common holly are imitated in some of the evergreen oaks. Some of these latter, especially varieties of the common *Quercus ilex*, are very like the olive. The trifoliate leaves of the wood-sorrel are very similar to those of the white clover, and both are represented, except that the leaves are four-lobed instead of three, in the cryptogamic genus Marsilea. There is an antarctic species of *Caltha*, or marsh marigold, in which the leaves strongly remind us of the Venus's fly-trap, and this originated its specific name (*Caltha dionæfolia*.) The digitate leaves of some of the cultivated species of *Aralia* might easily be mistaken for those of the castor-oil plant,

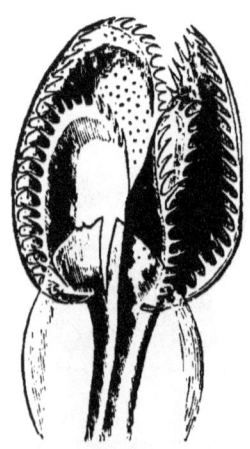

Fig. 70.—Leaf of *Caltha dionæfolia*.

although the former is allied to the ivy, and the latter to the far-distant *Euphorbiaceæ*. Another striking likeness in cultivated plants may be found in two variegated leaved species, with bright crimson veins. These are an Acanthaceous plant (*Gymnostachyum Verschaffeltii*),[1] and one of the Apocynaceæ (*Echites rubrovenosa*).[2] The size, form, colour, and mode of venation is almost identical. Numerous examples of pairs of plants resembling each other, chiefly in foliage, have been exhibited at the meetings of the Linnæan and other scientific societies.[3]

The inflorescence sometimes has a puzzling resemblance in one plant, or series of plants, to others with which they have no natural affinity. Some of the large African species of *Polygala* might easily be mistaken for *Papilionaceæ*. So again the Fig-marigolds (*Mesembryanthemum*) have a general likeness to the flowers of composite plants. A more extraordinary instance is in a genus of umbelliferous plants, of which two Australian species (*Actinotus*)

Fig. 71.—*Actinotus*.

---

[1] Figured in "Flora de Serres," pl. 1,581. [2] Ibid., pl. 1,728.
[3] For lists see "Nature," May 26, 1870, and May 4, 1871.

are strangely like ox-eye daisies (fig. 71). Our pretty little yellow "rock rose" (*Helianthemum*) reminds one of the yellow species of *Potentilla*, or the wood crowfoot (*Ranunculus*), and yet all three flowers belong to widely-separated natural orders. An instance also occurs to us in which an experienced botanist misnamed the flowers of *Coffea bengalensis*, as those of *Tabernemontana*, although they belong to families with no family connection. So also the inflorescence of *Dodecatheon*, nearly allied to the Cyclamen of the gardens, is not uncommonly mistaken for that of the dog's tooth violet (*Erythronium dens canis*), which it imitates in size, form, colour, and even in the bending backwards of the petals. Many of the myrtle family are excellent imitators of the *Rosaceæ*, and the rotate flowers of the

Fig. 72.—Rock Rose (*Helianthemum*).

fragrant white jasmins of India remind one strongly of *Apocynaceæ*.

The most striking instances of recurrence of type will be found amongst fruits, and perhaps the most numerous. Before "mimicry" was thought of in animals or plants, it had been remarked, as a singular coincidence, that the seeds of an Indian tree, *Mesua ferrea*, were like chestnuts. These seeds are not only alike in size, form, and colour, but also in character, so that they are eaten as a dessert fruit, in a similar manner. The Indian tree belongs to

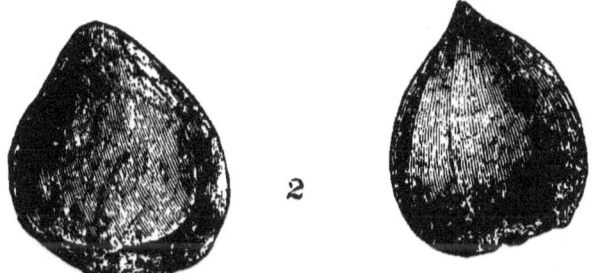

Fig. 73.—Seeds of *Mesua ferrea*, natural size.

the same family as the gamboge and mangosteen, whilst the chestnut finds a place with the oak, in a family far removed. Somewhat alike to these, but less striking, is the seed of the horse chestnut, the fancied resemblance being perpetuated in the name.

There is also great similarity between some of what are termed indehiscent legumes of the *Leguminosæ*, and drupes of the *Rosaceæ*; as, for instance, the pod of the tonquin bean and the fruit of an almond. The families are too closely allied, however, to give much weight to these resemblances. Perhaps the pod-like fruits of some of the caper family (*Capparidaceæ*) and their similarity to those of some of the Leguminosæ is more noteworthy. The form, size, and colour of some small gourds, of the cucumber family, such as the colocynth and the orange gourd, approximate to the fruit of the orange.

The winged fruits of the maples, with the seed at one extremity and a veined wing at the other, is a type of "samara" which is found repeated again in other families. It occurs in a genus of *Polygalaceæ*, which is found chiefly in tropical South America. Our figure is *Securidaca tomentosa* (fig. 74). The

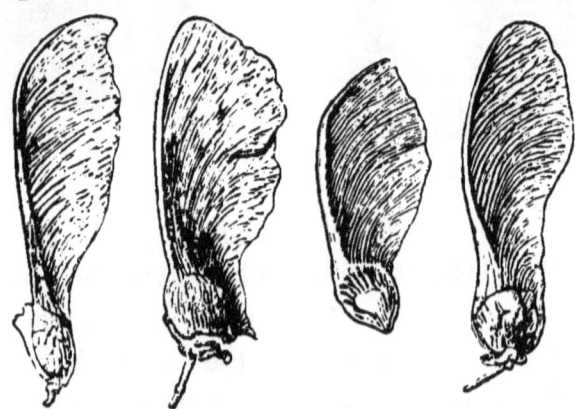

Fig. 74.—Samara of *Securidaca tomentosa, Heteropterys laurifolia, Gallesia gorancma, Seguiera floribunda.*

same form is found again in *Malpighiaceæ*, of which

the species are mostly South American. This is represented by *Heteropterys laurifolia* (fig. 74) ; and yet again in the *Phytolaccaceæ*, the same kind of samara is found in at least two genera, of which we have illustrated *Gallesia gorancma* and *Seguiera floribunda*. These four illustrations are from three natural orders, all separate from each other and from the maples, and yet, not only is the size and form the same, but also the veining in the wings. So deceptive is the resemblance between these fruits, that only dissection and analysis could determine one from the other.

Another type of samara is that of the elm, in which the seed occupies the centre, surrounded by a wing. Our common forms are those of the common elm (*Ulmus campestris*) and the wych elm (*Ulmus montana*), the latter being the largest. This form of fruit is imitated in *Ptelea trifoliata*, a tree of the *Rutaceæ*, and in a species of *Hiræa*, one of the *Malpighiaceæ*.

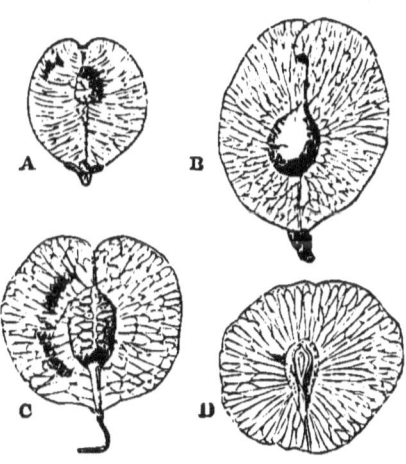

Fig. 75.—Samara of (*a*) *Ulmus campestris*, (*b*) *Ulmus montana*, (*c*) *Ptelea trifoliata*, (*d*) *Hiræa*.

In like manner there is more than a merely superficial resemblance, but almost identity, although the families to which the trees belong have no close relationship (fig. 75).

Winged seeds, as distinguished from the winged

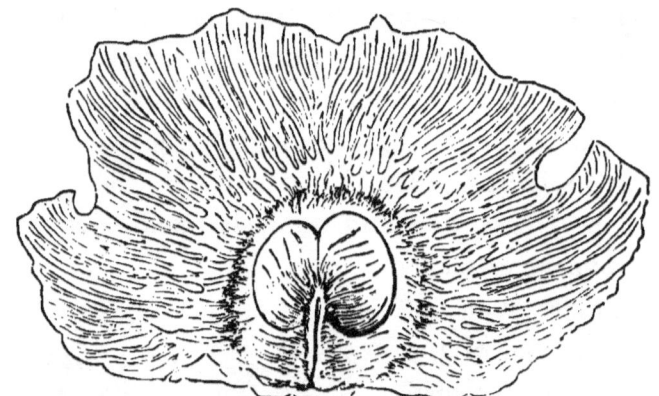

Fig. 76.—Seed of *Calosanthes indica.*

Fig. 77.—Seed of *Zanonia macrocarpa.*

fruits just alluded to, are highly developed, and of considerable size in the family of trumpet flowers

(*Bignoniaceæ*). One of the best-known forms (*Calosanthes indica*) is given in the woodcut (fig. 76). The membrane which surrounds the seed is beautifully delicate and transparent, and is a favourite object with microscopists. This type of seed is represented again in the cucumber family, in which winged seeds are rare, and is, in fact, almost an imitation of the seed of the *Calosanthes*. In our figure (fig. 76) it has been reduced by about one-third, so as to bring it within limits of the page. It does not differ more from the seed of one of the Bignoniaceæ than these seeds differ amongst themselves. In another family (*Apocynaceæ*), similar winged seeds occur (as in *Aspidosperma excelsum* from Guatemala), although it is not a special feature in that family for the seeds to be expanded in a membranaceous wing.

Every schoolboy is acquainted with the downy crest of the achenes, or fruits, of the dandelion and thistle. This crest of delicate hairs, or pappus, is common in composite plants, but it is not confined to them. From the annexed woodcut (fig. 78) it will be seen that one of the forms, with the crest sessile, is reproduced in three other families, viz., in *Sarcostemma* (*a*) one of the *Asclepiadaceæ*, in *Echites scabra*, one of the *Apocynaceæ* (*b*), in the willow herb or *Epilobium*, one of the *Onagraceæ* (*c*), and in the milk thistle, *Silybum marianum*, one of the *Compositæ*. There is more difference in the character of the seeds themselves than in the crest which surmounts them.

Seeds are also in some instances invested with a dense down, as in the familiar example of the cotton

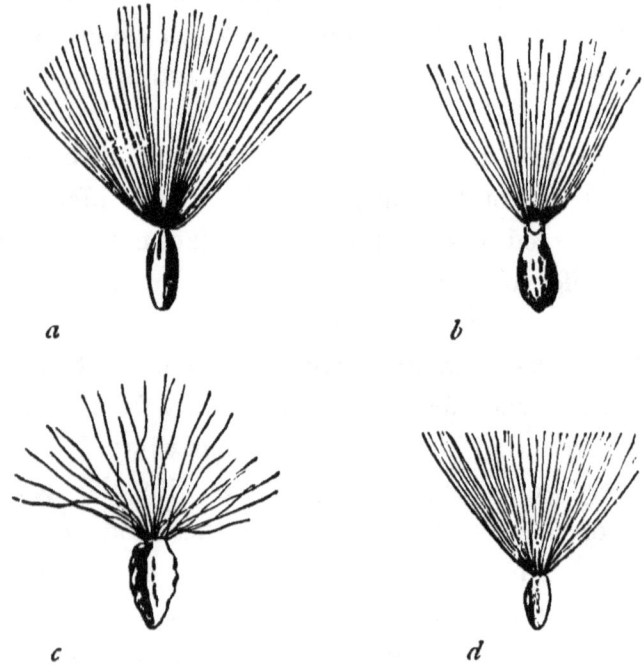

Fig. 78.—(*a*) Crested seed of *Sarcostemma*, (*b*) *Echites scabra*, (*c*) Willow-herb (*Epilobium*), (*d*) Milk Thistle (*Silybum marianum*).

plant, in which the seed is covered with the substance called "cotton." In some other plants, which are closely allied, this cotton is silky and shining (as in *Bombax* and *Eriodendron*), but this same silky substance invests the seeds in some of the *Asclepiadaceæ*, which family is far removed from the above, and also from the *Convolvulaceæ*, in which latter family there are instances of *Ipomæa*, the seeds of which are invested in one species with a substance like "silk-cotton," and in another with a substance resembling a native Indian cotton, approaching to the texture of wool.

It would not be difficult to go on repeating instances of such coincidences in fruits and seeds for many pages, but we must rest content with another series. Most persons are aware that the prevalent type of seed in the fir tree, or coniferous family, consists in a brown hard seed, with an extension of membrane into a kind of wing. Of course, there are different forms of pine seeds, some of which have no wing; but we desire to indicate that the winged type is repeated in other families, not at all related to the *Coniferæ*. This is the case with the seeds of some of the species of *Lagerstræmia*, which are allied to our "purple loosestrife;" also in *Cedrela*, and some other genera in the natural order to which the mahogany tree belongs; and, to some extent, in *Ailanthus*, wherein there is an approach to the samara of the ash, the winged fruits being, in fact, "samara," and not seeds. Fir cones themselves are almost imitated in the fruits of some of the palms, in so far as appearance goes, although in structure very different, and the scales are imbricated in the opposite direction; nevertheless they might pass for an imitation.

If we go still deeper into the structure of plants and investigate their secretions, we shall encounter here and there coincidences of strange significance. In the lettuce is repeated the narcotic milky juice of the poppy. In the American *Loasaceæ* the stinging properties of the nettles. In the figs (*Ficus*) of India, the rubber trees of Para (*Siphonia*) and the Urceolas of Asia, we have in three different families, and to a certain extent in some others, plants furnishing the same kind of milky juice which consolidates into caoutchouc, or india-rubber. The acrid juice which is secreted by some of the *Anacardiaceæ* has its analogue in the *Euphorbiaceæ*. The smoke of the wood of *Excœcaria agallocha* when burnt is said to affect the eyes with intolerable pain; and so also the manchineel, which belongs to the same family, and that of another tree, referred to the *Anacardiaceæ*. The native in Brazil poisons his arrows with the juice of the mandioca plant, which belongs to the Euphorbiaceæ; the Fiji Islanders with that of an *Antiaris*, which is of the bread fruit family; and on the Orinoco the famous curare is obtained from a *Strychnos*. The ancient Briton obtained his blue dye from the woad (*Isatis tinctoria*), the Hindoo, and other Asiatics, from the indigo plant, and the

Assamese from a *Ruellia*. We do not pretend to assert that these are all instances of mimetic resemblance, or that these, and scores of similar coincidences, would give any support to a theory of natural selection and survival of the fittest. All that we are justified in proposing is, that these circumstances should be borne in remembrance in connection with the strange coincidences of form to which we have devoted the preceding pages.

Amongst fungi there are also striking resemblances, which have been detailed more fully in another place.[1] Special attention was first directed to this subject by Mr. Worthington Smith, who gave several rather striking examples, although the pairs are more closely allied than those selected amongst the flowering plants. Thus, one poisonous species, *Agaricus* (*hebeloma*) *fastibilis*, greatly resembling the edible mushroom, *Agaricus* (*psalliota*) *campestris*, came up in great numbers upon a mushroom bed, and might have caused a disastrous result, had not the fact been detected by an adept. Another instance was also that of a mass of fungi which made their appearance on a mushroom bed. At first sight these closely resembled the variety of an edible species which not unusually comes up in clusters on old beds; it has white spores, with a lobed and undulated white

---

[1] "Mimicry in Fungi, in Grevillea," vol. ix., p. 151.

pileus, *Agaricus* (*clitocybe*) *dealbatus*. The imitating fungus had the same wavy cap, white colour, and fungoid odour, but the spores were pink, and its structural features were distinctly those of quite a different species, *Agaricus* (*clitopilus*) *orcella*. In this instance both were quite innocuous. Two wholly distinct, but very similar fungi commonly grow together on wood ashes, or scorched places, where charcoal has been burnt; these are *Cantharellus carbonarius* and *Agaricus* (*collybia*) *atratus*. Then, again, another pair of fungi, in which sulphur colour prevails, are found growing together on wood. These are *Agaricus* (*hypholoma*) *fascicularis* and *Agaricus* (*flammula*) *conissans*, or, similarly *Agaricus* (*hypholoma*) *capnoides* and *Agaricus* (*flammula*) *alnicola*. In all these four the coincidence of colour, form, size, mode of growth, and even habitat, is complete. With any one of these, again, may be compared the recently discovered *Agaricus* (*clitocybe*) *Sadleri*, which has white spores. Here we have five yellow species found growing on wood, and so like each other that an ordinary observer would consider them all as the same species, not taking the colour of the spores into account. There is, moreover, a small agaric, which is known to the majority of mycologists on account of its strong odour of stinking fish (*Agaricus cucumis*). It grows on the ground, and upon fragments of wood, and has red-brown spores. Yet there is an imitator in a small fungus with white spores, found in just the same localities, with the identical fishy odour. According to all authority and experience the difference in the colour of the spores is not a mere difference of species, but indicates quite a separate and distinct group of species.

We might also indicate as further removed from each other such species as *Agaricus (tricholoma) nudus*, a handsome violet species, which when well grown is scarce to be distinguished from *Cortinarius violaceus*, except that in the former the spores are white, and in the latter rusty.

Taking a still wider range we encounter equally startling resemblances between widely separated groups, such as the whole hypogœous *Gasteromycetes*, which in form, size, odour, habit, and all save fructification imitate the truffles (*Tuberaceæ*). Or, opposing certain genera we have in *Podaxon* a resemblance to *Coprinus*, and *Hypolyssus* might be mistaken for an immature *Crucibulum*. The larger species of *Peziza* sometimes approach in habit *Craterellus*. And in *Cyphella*, with its naked spores, every feature besides corresponds with the small *Pezizæ*, some being like the section *Hymenoscypha*, others that of *Dasyscypha*, and others *Mollisia*.

Comparing fungi with other cryptogamia, the gelatinous species of *Tremella* are just like such algæ as *Nostoc*. In lichens the species of *Lecidea* approximate so closely that only experts can distinguish them from *Patellaria* amongst fungi. *Bæomyces* amongst lichens imitates *Stilbum* in fungi, whilst the graphideous lichens seem to coalesce almost with *Hysterium*, and *Platygrapha* with *Stictis*.

Already our comparisons are too technical, and we must rest content with thus much allusion to a subject which presupposes too much practical knowledge for a popular volume. We may, nevertheless, urge that amongst the lower order of plants there are coincidences as striking as those instanced in flowering plants. Whatever the interpretation may be, the facts are worthy of remembrance, since we may hereafter, subject to a wider experience, suggest reasons which would now be regarded as premature.

In bringing this interesting subject to a conclusion we may briefly allude to certain fancied resemblances which are occasionally met with, reminding us strongly of members of the animal kingdom. Certain fruits and seeds are supposed to resemble beetles, bugs, &c., and some flowers to mimic bees, flies, and butterflies. In passing we have alluded to some of these, and shall now rest content with reference to the snake nut of Demerara. This fruit was discovered and made known by Sir Robert Schomburgk in 1840.[1] "For several years past," he says, "nuts of the size

---

[1] "Annals of Natural History," 1840, vol., p. 202.

of a walnut were brought down from the interior to Georgetown in Demerara, the kernel of which, when opened, and the membrane which covered it being removed, displayed the striking resemblance to a snake coiled up. There was the head, the mouth, the eyes so complete, that one unacquainted with the fact would have believed them to be an imitation made by human hands, and not a freak of nature. As is often the case with the productions of the interior, the colonists were entirely unacquainted with the mode of growth of the plant which produced these strange nuts. They were generally found after the annual swelling of the Essequibo had subsided along its banks, and for a length of time it was pretended that they grew on a creeper, and from the resemblance of its kernel to a snake it was supposed that it might prove an antidote to snake poison." Subsequently it was found to be the produce of a large tree (*Ophiocaryon serpentinum*) belonging to the same family as the horse chestnut. Our figure represents a nut cut open, and the kernel exposed (fig. 79).

Fig. 79.— Snake nut (*Ophiocaryon serpentinum*) cut open.

As in some sort to counterbalance a too rigid

application of utilitarianism to forms and modifications of plant structure, especially such as relate to the subject of this chapter, we shall end with a quotation from Mr. A. W. Bennett. He says, " I cannot myself get away from the conclusion that we must attribute the tendency to variation which is admitted to be the material on which natural selection works, to some inherent force belonging of necessity to the functions of life, whether animal or vegetable, which is independent of, and in some sense superior to, the forces that govern the inorganic world. Above all, we are compelled to recur to the pre-Darwinian doctrine of Design ; and to believe that nature has some general purpose in the different modes in which life is manifested, a purpose not in all cases for the immediate advantage of the individual species, but in furtherance of some design of general harmony which it may take centuries of unwearied observation and laborious toil before we discover the key by which we may be able to unlock it.[1]

---

[1] A. W. Bennett on "Mimicry in Plants" in "Popular Science Review," vol. xi. (1872), p. 10.

CHAPTER XVI.

GIANTS.

"THERE were giants in those days" scarcely includes those of the vegetable world, for the facts which relate to the most gigantic of plant productions are of recent date. Under the term " giant " we do not purpose to include unusual developments of *individuals*, but to refer to *species*, of which large dimensions is an attribute. Large oaks, large elms, large forest trees of various kinds are enumerated in all books of forestry, and these have their own interest, but not the same interest as that which attaches to plants which are normally of extraordinary size. Literally, then, we have to deal with vegetable Titans, with "mammoth" trees, and gigantic flowers, commonly attaining dimensions far in excess of ordinary trees and flowers; their claim to notice being their normally unusual size.

It was for some time supposed that the largest of all known trees were the conifers of the western side of the North American continent. The trees known to Englishmen as *Wellingtonia* and to Americans as *Sequoia* were, up to a recent date, regarded as the mammoth trees. This supremacy is now broken down in favour of the "big trees" of Australia, although it must be confessed that it is very difficult to determine what are the reliable dimensions of trees recorded in both countries.

In a work of authority[1] it is said that the big trees (*Sequoia gigantea*) extend along a line of two hundred and forty miles, and moreover, that the highest yet discovered, which is in the Calaveras Grove, is three hundred and twenty-five feet. The grizzly giant of the Mariposi Grove is ninety-three feet in circumference at the ground. These dimensions have been greatly exceeded *by report*, but the sensational heights of four hundred feet and upwards are believed to be wholly unreliable. Dr. C. F. Winslow, in "The California Farmer," has written that "the trees of very large dimensions number considerably more than one hundred. Mr. Blake measured one ninety-four feet in circumference at the root, the side of which had been partly burnt by contact with another tree, the head of which had fallen against it. The latter can be measured four hundred and fifty feet from its head to its root. A large portion of this fallen monster is still to be seen and examined; and, by the measurement of Mr. Lapham, it is said to be ten

---

[1] Watson's "Botany of California," vol. ii. p. 117.

feet in diameter at three hundred and fifty feet from its upturned root."[1]

Dr. Berthold Seemann,[2] a most trustworthy observer, has given a detailed account of some of the most remarkable of the sequoias, in which the "Hercules" is named as three hundred and twenty-five feet high, and ninety-seven in circumference at the base. "Uncle Tom's Cabin" claims to be three hundred feet high, and seventy-five feet in circumference. The "big tree," which was felled, was ninety-six feet in circumference at the base, and solid throughout. This was effected by boring holes with augers, and then connecting them by means of an axe. Twenty-five men were thus occupied for five days. When this was done, it was only by applying a wedge and strong leverage, favoured by a heavy breeze, that the overthrow was accomplished; stones and earth being cast up with such force that these records of the fall may be seen on surrounding trees, to the height of nearly a hundred feet. Although we have sought, and enquired diligently, we do not find reliable grounds for rejecting Sereno Watson's maximum height of three hundred and twenty-five feet.

The gigantic trees of Australia are gum trees, a

---

[1] Hooker's "Kew Garden Miscellany," vol. xii. (1855), p. 27.
[2] Dr. B. Seemann, "Ann. Nat. Hist.," March, 1859.

species of eucalyptus, for the details of the dimensions of which we are indebted to Baron F. von Mueller. Of later years, as easier tracks have been opened, increased heights have been ascertained. "The highest tree previously known was a Karri-eucalyptus (*Eucalyptus colossea*) in one of the glens of the Warren River of Western Australia, where it rises to approximately four hundred feet high. Into the hollow trunk of this Karri three riders, with an additional pack horse, could enter and turn without dismounting. Mr. D. Boyle measured a fallen tree (*Eucalyptus amygdalina*), in the deep recesses of Dandenong, and obtained for it the length of four hundred and twenty feet, with proportionate width; while Mr. G. Klein took the measurement of an eucalyptus on the Black Spur, ten miles distant from Healesville, four hundred and eighty feet high."[1] Mr. G. Robinson estimated an eucalyptus in the black ranges of Berwick at five hundred feet. "It is not at all likely that in these isolated enquiries chance has led to the really highest trees, which the most secluded and the least accessible spots may still conceal. It seems, however, almost beyond dispute that the trees of Australia rival in length, though evidently not in thickness, even the renowned forest giants of California (*Sequoia gigantea*). We

---

[1] Cooper's "Forest Culture," p. 198.

possess a standard of comparison in the spire of the cathedral of Strasburg, the highest of any cathedral of the globe, which sends its lofty pinnacle to the height of four hundred and forty-six feet; or in the great pyramids of Cheops, four hundred and eighty feet high, which, if raised in our ranges, would be overshadowed probably by Eucalyptus trees."

In one sense a giant, and in another sense a dwarf, there is no more remarkable plant to be found than that called after the name of its discoverer, Dr. Welwitsch (*Welwitschia mirabilis*). "Several miles before reaching Cape Negro the coast rises to a height of about 300 or 400 feet, forming a continuous plateau, extending over six miles inland, as flat as a table." Amongst the vegetation of this plateau a dwarf tree was particularly remarkable. This, "with a diameter of stem often of four feet, never rose higher above the surface than one foot, and which, through its entire duration that not unfrequently might exceed a century, always retained the two woody leaves which it threw up at the time of germination, and besides these it never puts forth another. The entire plant looks like a round table, a foot high, projecting over the tolerably hard sandy soil; the two opposite leaves (often a fathom long by two to two and a half feet broad) extend on the soil to its margin, each of them split up into numerous ribbon-like segments." The flowers of this singular plant are produced in clusters, and have the form of crimson cones, not unlike those of the larch.[1]

The first announcement of such a singular "freak of plant-life" was received with some incredulity, but when not only drawings but the plants themselves arrived, the incredulity became changed to astonishment; and its whole history, unfolded in a most complete and thoroughly illustrated memoir,[2] by Sir Joseph Hooker, passed into the records of science as one of the most remarkable discoveries of plant-life which the present century has been able to produce.

The tree which attains the greatest lateral expansion is the Indian fig, or banyan, which drops down rope-like shoots from their branches, and these, when they reach the soil, enter it and take root, thus becoming in the course of time subsidiary trunks. The increase in this manner might also be supposed to be indefinite, by the addition of new trunks, as the branches extend themselves. Milton has alluded to this tree as—

> The fig-tree; not that kind for fruit renowned;
> But such as at this day, to Indians known
> In Malabar or Deccan, spreads her arms,
> Branching so broad and long, that in the ground
> The bended twigs take root, and daughters grow
> About the mother tree, a pillar'd shade
> High overarch'd, and echoing walks between.

---

[1] Welwitsch, "West African Botany." "Journ. Linn. Society," v. p. 185. [2] Hooker in "Linnæan Transactions."

The great tree of the Nerbudda, often alluded to as the most important of these trees, covers a very large area, of which a circumference of two thousand feet is still remaining, though some has been swept away. Three hundred and twenty main trunks have been counted, while there are smaller ones to the number of some three thousand, and each of these is constantly sending forth branches and forming pendent root-stocks so as to extend and increase the colony. "Immense popular assemblies are sometimes convened beneath this patriarchal fig, and it has been known to shelter seven thousand men at one time beneath its ample shadow."[1]

The largest forms of the strange cactus tribe are found in California and Mexico. Missionaries who visited these regions more than a century ago mention them as remarkable trees without leaves, but branched, and sixty feet in height. The giant cactus (*Cereus giganteus*) inhabits the wildest and most inhospitable regions, where "its fleshy shoots will strike root and grow to a surprising size, in chasms in heaps of stones, where the closest examination can scarcely discover a particle of vegetable soil. Its form is various, and mostly dependent on its age; the first shape it assumes is that of an immense club, standing upright in the ground, and of double the

---

[1] Forbes's "Oriental Memoirs."

circumference of the lower part at the top. This form is very striking while the plant is still only from two to six feet high, but as it grows taller the thickness becomes more equal, and when it attains the height of twenty-five feet it looks like a regular pillar; after this it begins to throw out its branches. These come out at first in a globular shape, but turn upward as they elongate, and then grow parallel to the trunk, and at a certain distance from it, so that a Cereus with many branches looks like an immense candelabrum, especially as the branches are mostly symmetrically arranged round the trunk, of which the diameter is not usually more than a foot and a half, or rarely a foot more."[1] They vary much in height; some are said to be thirty-six or forty feet, and others not less than sixty. "As seen rising from the extreme point of a rock, where a surface of a few inches square forms their sole support, one cannot help wondering that the first storm does not tear them from their airy elevation." "Wonderful as each plant is, when regarded singly, as a grand specimen of vegetable life, these solemn, silent forms which stand motionless, even in a hurricane, give a somewhat dreary character to the landscape. Some look like petrified giants, stretching out their arms in speechless pain, and others stand like lonely sentinels,

---

[1] Möllhausen's "Journey to the Pacific," ii. p. 248.

keeping their watch on the edge of precipices and gazing into the abyss."

We who are accustomed to see such climbing plants in our woods as the honeysuckle and hop, have but a poor conception of what climbing plants become in a tropical forest. Kingsley alludes to a magnificent wild vine or liantasse (*Schnella excisa*), "so grand that its form strikes even the negro and the Indian. You see that at once by the form of its cable—six or eight inches across in one direction and three or four in another, furbelowed all down the middle into regular knots, and looking like a chain cable between two flexible iron bars. At another of the loops, about as thick as your arm, your companion, if you have a forester with you, will spring joyfully. With a few blows of his cutlass he will sever it as high up as he can reach, and again below some three feet down; and while you are wondering at this seemingly wanton destruction he lifts the bar on high, throws his head back, and pours down his thirsty throat a pint or more of pure cold water. This hidden treasure is, strange as it may seem, the ascending sap, or rather the ascending pure rain water, which has been taken up by the roots, and is hurrying aloft to be elaborated into sap, and leaf, and flower, and fruit, and fresh tissue for the very stem up which it originally climbed; and therefore it is that the woodman cuts the water-vine through first at the top of the piece which he wants, and not at the bottom; for so rapid is the ascent of the sap that if he cut the stem below, the water would have all fled upwards before he could cut it off above. Meanwhile the old story of Jack and the Bean-stalk comes into your mind."[1] Such a "bean-stalk" must be that of *Entada scandens*, a tropical climber of the bean family, which has pods nearly two yards long and five inches broad, with beans as large as the flattened "Normandy pippins," so often seen in the grocers' windows.

Rattans, which are the terror of schoolboys, are also the dread of the traveller, but for different reasons. These palms, often with stems not thicker than the little finger, are armed with rigid pointed spines, climbing by their aid to the tops of the highest trees, then dropping their extremities to the ground, and rising again until they will attain a length of several hundred feet. In the bulk of stem they are diminutive, but in extension are worthy of note as "giants." They are abundant in all the forests of the Malay and Philippine archipelago, and are everywhere extensively used as cordage, or for the manufacture of basket work. "These singular plants creep along the ground, or climb trees, and, according to the species, to the length of from one

---

[1] Kingsley's "At Last," p. 159.

hundred to twelve hundred feet."[1] The latter length is given on the authority of Rumphius, but it is very difficult to obtain authentic records of the length to which they will attain. It is not uncommon for the ordinary species, the common "canes" which form an article of commerce, to reach lengths varying from three to five hundred feet, and yet with but little increase of thickness through the entire length. These climbing palms contribute much to produce that character of impenetrable thicket which is so peculiar to tropical forests.

What, after all, are the bamboos but gigantic grasses. They belong to the same family, and possess the family likeness, growing in dense tufts, or tussocks, with seeds resembling those of oats. They are natives of tropical countries, where their uses are manifold. "The bamboo, full grown, forms usually a more or less developed stock, sometimes up to three feet high, formed chiefly of old trunks of the dead haulms and an entanglement of roots, from which ten to fifty, and even up to a hundred haulms arise of the thickness of one's arm to that of the human thigh, often attaining upwards of one hundred and twenty feet in height."[2]

The rapidity of the growth of bamboo shoots has

---

[1] Crawfurd's "Dictionary of Indian Archipelago," p. 364.
[2] Kurz, "The Bamboo and its Use," p. 242.

often been alluded to. The usual period during which they attain their full height varies between two and three months. A bamboo in a hothouse in Glasgow was seen to grow one foot in twenty-four hours. Mr. Fortune made various measurements of the growth of bamboos in the Chinese jungles, and has reported the growth to have been from two to two and a half feet in twenty-four hours, with the greatest growth during the night. The culms, or stems, are hollow, like a reed, with joints at regular distances, so that, except for size, they would be accepted as reeds. Cut off at the joints they are convertible into kitchen utensils, some being large enough for pails; and when pierced through at the joints, so as to form continuous pipes, they are employed as aqueducts. Only those who have visited India, China, or Malayan countries could imagine the innumerable uses to which these gigantic grasses are applied.

Palms are tropical trees of a peculiar growth, having usually a single erect stem without branches, only one or two species ever producing a branch. In appearance, with their large expanded fronds, or leaves, they have but little in common with ordinary trees. Some of the palms attain to a considerable size, although not comparable with the big trees of California or Australia, yet not less remarkable when their structure is taken into account. It is, however, the leaves to which we would allude as especially worthy of notice here. We may have the pinnate, or feathery leaf, similar to an ordinary fern frond, and the fan-shaped leaf. Of the former, the Jupati, one of the Brazilian palms (*Raphia tædigera*), Wallace says, "Its comparatively short stem enables us to fully appreciate the enormous size of the leaves, which are at the same time equally remarkable for their elegant form. They rise nearly vertically from the stem, and bend out on every side in graceful curves, forming a magnificent plume seventy feet in height, and forty in diameter. I have cut down and measured leaves forty-eight and fifty feet long, but could never get the largest."[1] Of another palm he writes (*Maximiliana regia*) : "The leaves of this tree are truly gigantic. I have measured specimens which have been cut by the Indians fifty feet long; and these did not contain the entire petiole, nor were they of the largest size."[2] Of the fan palms the most magnificent are the leaves of the Talipat palm (*Corypha umbraculifera*) of Ceylon, which are used as umbrellas and for tents, a large one being sufficient to cover and protect fifteen persons from the sun and rain. In making tents two or three leaves are usually sewn together.

Periodically the botanical world has been astonished by the report of some newly-discovered

---

[1] Wallace, "Palms of the Amazon," p. 43. [2] Ibid., p. 121.

giant. At one time it was the great *Rafflesia*, then the royal water-lily, and last, but not least, the monster arum of Beccari. The one solitary example of this family which belongs to our climate is the little "wake-robin," or "lords and ladies" of our hedgerows. In the centre of the tuft of glossy leaves rises the singular flower, or what is commonly designated as the flower, but which really is a large colony of minute flowers, surrounding the base of an erect club-shaped column called a spadix, and enclosed in a sheath or envelope, rising to a sharp point and opening on one side so as to expose but a glimpse of the column within. The root is a small tuber, or corm, containing a quantity of starch, which, during the time of Queen Elizabeth, was collected for starching the "ruffles" of the court. Just such a plant, on an enlarged scale, was discovered by the Italian botanist in Sumatra. The tuber in this species was five feet in circumference. The leaves, on foot stalks ten feet in length, were much divided, and covered an area of forty-five feet in circumference.[1] The spadix, or central column, was nearly six feet in height. The diameter of the spathe was nearly three feet, of a bell shape, with crumpled and deeply-toothed edges, of a pale greenish colour within, and externally of a bright blackish purple.

---

[1] "Gardener's Chronicle," vol. x. (1878), p. 788.

In the accompanying woodcut, the central spadix rising out of the bell-shaped cup should be near six feet, so that the figure is reduced to one twenty-fifth of the height of the original, which has been named *Amorphophallus Titanum*.

The monarch of flowers, in respect to size, is that first discovered by Sir Stamford Raffles, and named after him, Rafflesia. It is a large fleshy parasite, growing on the roots of other plants, without leaves, and consisting entirely of a single enormous flower, "of a very thick substance, the petals and nectary being but in a few places less than a quarter of an inch

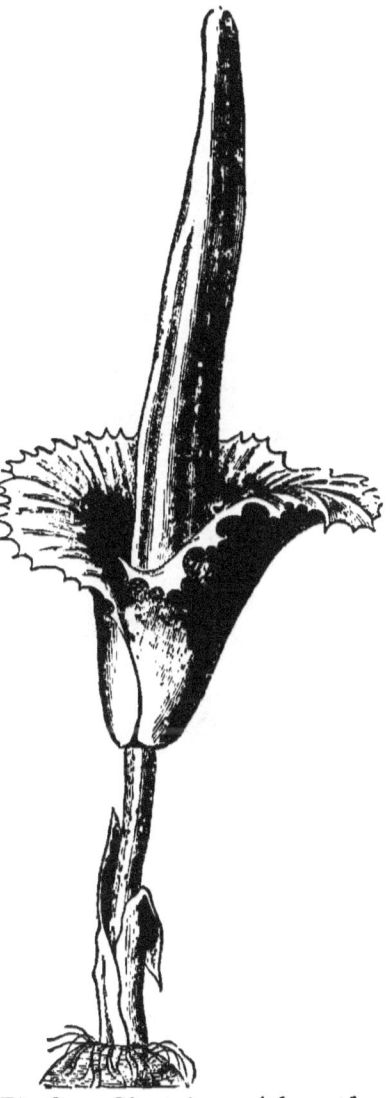

Fig. 80. — Giant Arum (*Amorphophallus Titanum*) greatly reduced.

thick, and in some places three quarters of an inch: the substance of it was very succulent. When I first saw it, a swarm of flies were hovering over the mouth of the nectary, and apparently laying their eggs in the substance of it. It had precisely the smell of tainted beef. It measured a full yard

Fig. 81.—*Rafflesia Arnoldi*, reduced from photograph of living flower.

across; the petals, which were subrotund, being twelve inches from the base to the apex, and it being about a foot from the insertion of the one petal to the opposite one. The nectary, in the opinion of all of us, would hold twelve pints, and the weight of this prodigy we calculated to be fifteen pounds."[1]

The flower was first discovered in 1818, on the Manna River in Sumatra, where it is said to be known by the name of the "Devil's Siri box!" Dr. Arnold says that when he first saw it in the jungle it made a powerful impression on him. "To tell the truth, had I been alone, and had there been no witnesses, I should, I think, have been fearful of mentioning the dimensions of this flower, so much does it exceed every flower I have ever seen or heard of." Another species has been found in Java, but not quite of such an enormous size.

Second in size are the flowers of one of the birthworts, climbing aristolochias of tropical forests. Humboldt gave the first intimation of the existence of these giants in these words: "On the shady banks of the Magdalena River, in South America, grows a climbing aristolochia, whose blossoms, measuring four feet in circumference, the Indian children sportively draw on their heads as caps."[1] This species (*Aristolochia grandiflora*), or what is believed to be the same species, is called "pelican flower" in the West Indies, from the resemblance of its young and

---

[1] Hooker's "Companion to Botanical Magazine," i. (1835), p. 262. "Transactions of Linnæan Society," vol. xiii.

[2] Humboldt, "Views of Nature" (1850), p. 230. "Botanical Magazine," pl. 4,368.

unopened flower to the head of a pelican at rest. Miers states that he had often seen it in Brazil, where he was led to compare the large flaccid blossoms on the bushes with coloured pocket-handkerchiefs laid out to dry. Lunan remarks that the odour is so abominably fœtid that it is detested and shunned by most animals; and when hogs venture, through necessity, to eat of it, it destroys them.[1] Tussac, noting the same plant in the Antilles, says that a whole herd of swine, having been driven into the woods where this plant was common, had entirely perished from eating the roots and young stems. Another species, which has now flowered two or three times in this country (*Aristolochia Goldicana*), comes from Old Calabar River and Sierra Leone.[2] The flowers reach to twenty-six inches in length and eleven inches in diameter at the mouth, when grown here. Like the other, it has a strong and powerful odour as of putrid meat. Our figure of this species is very considerably reduced, but it represents the form, and from the measurements of its diameter

---

[1] "Transactions Linnæean Society," xxv., p. 185, pl. 14.

[2] "The largest flowers in the world, besides those belonging to the *Compositæ* (the Mexican *Helianthus annuus*), are produced by *Rafflesia Arnoldi*, *Aristolochia*, *Datura*, *Barringtonia*, *Gustavia*, *Carolinea*, *Lecythis*, *Nymphæa*, *Nelumbium*, *Victoria regia*, *Magnolia*, *Cactus*, the *Orchideæ*, and the Liliaceous forms."—Humboldt, "Views of Nature," p. 348.

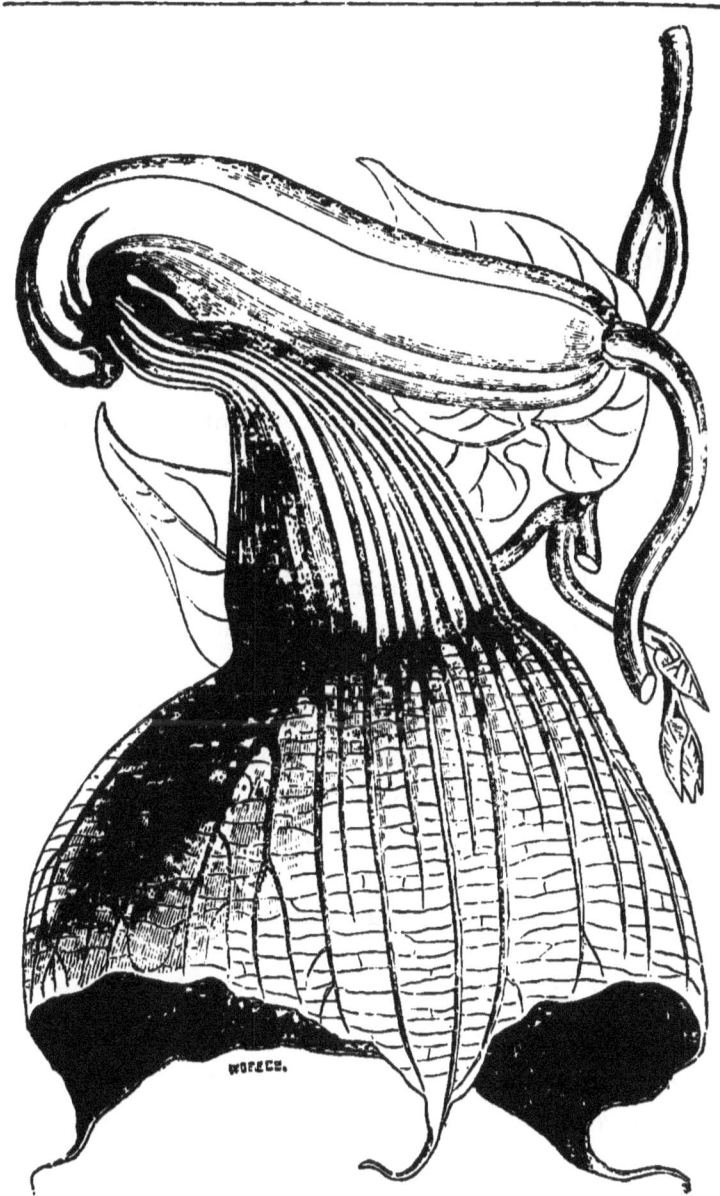

Fig. 82.—Flower of *Aristolochia Goldieana* reduced.

at the funnel-like mouth, it must be conceded that it is no exaggeration to say that it may be placed like a cap on the head of a very broad-headed adult.

The flowers of the night-blooming cereus (*Cereus grandiflorus*) are very different in character, and inferior in size; they have, however, the merit of possessing a very grateful fragrance. It is alluded to here as one of the largest of blossoms, attaining, it is said, when fully expanded, a diameter of a foot, but as this measurement is taken from tip to tip of the petals, it does not seem so large as a cup-shaped flower would be.

Amongst lilies there are two or three magnificent species which deserve remembrance. Such, for example, is *Lilium giganteum*, of which a dried stem is preserved in one of the museums at Kew. Let the imagination strive to picture a gorgeous white lily, with a flower stem eleven and a half inches in circumference at the base, and rising to a height of thirteen feet, bearing blossoms as large as tumbler glasses. It might be said literally that "Solomon in all his glory was not arrayed like one of these."

No allusion to extraordinary flowers would be considered as complete without reference to the royal water-lily (*Victoria regia*), dedicated to the Queen, and made the subject of two entire volumes, one on each side of the Atlantic, which, for size of page, are almost the largest of modern books. The oft-repeated account of its discovery on New Year's Day, 1837, by Sir Robert Schomburgk, whilst on his way up the River Berbice, has become a historic record, and is the basis of all detailed chronicles. "There were," he says, "gigantic leaves, five to six feet across, flat, with a broad rim, lighter green above, and vivid crimson below, floating upon the water; while in character with the wonderful foliage I saw luxuriant flowers, each consisting of numerous petals, passing in alternate tints, from pure white to rose and pink. The smooth water was covered with the blossoms, and as I rowed from one to the other I always found something new to admire. The flower-stalk is an inch thick near the calyx, and studded with elastic prickles about three quarters of an inch long. When expanded the four-leaved calyx measures a foot in diameter, but is concealed by the expansion of the hundred-petaled corolla. This beautiful flower, when it first unfolds, is white with a pink centre; the colour spreads as the bloom increases in age; and at a day old the whole is rose-coloured. As if to add to the charm of this noble water-lily, it diffuses a sweet scent. Ascending the river we found this plant frequently, and the higher we advanced the more gigantic did the specimens become; one leaf we measured was six feet five inches in diameter, the rim five and a half inches high, and the flowers a foot and a quarter across."[1]

If one were asked to determine the largest fruit hitherto known, it is probable that the answer must be some species of gourd or "pumpkin," the dried external portion of one such specimen being suspended in one of the museums of the royal gardens, Kew, with a diameter of about two feet. This far exceeds the largest "double cocoa-nut" (*Lodoicea seychellarum*) of which we have any experience. As far as we know, the full dimensions of the largest gourds have not been recorded, since they may attain, in their native and warmer climes, a much greater diameter than in cultivation.

If individual seeds are the subject of inquiry, then we are assured that the largest seeds of which we have hitherto any experience are the beans of a Mora tree (or as it is now called *Dimorphandra oleifera*) from Panama. These seeds are as much as six inches long, five inches broad, and four inches thick. If edible, such beans would not be requisite in any great numbers for an ordinary meal.

Justification might almost be found for an allusion to such large starchy roots as the elephant's foot, and yams of various species, in which great bulk is com-

---

[1] "Botanical Magazine," pl. 4,275 : "Annals of Natural History" (1838), p. 65.

bined with farinaceous qualities, which render them available, after the manner of gigantic potatoes, as articles of animal food.

Those truly elegant plants the Ferns, as popular as any of the members of the vegetable kingdom, have also their giants in the tree ferns of tropical climates. The "silver king" (*Cyathea dealbata*) has leaves, or fronds, from five to seven feet in length; and Dieffenbach found it growing in New Zealand with trunks upwards of forty-two feet in height. Another, which might be called the "monarch" (*Dicksonia antarctica*), has fronds from six to twelve feet in length, or more. One plant, cultivated in this country, and hence probably inferior in size to those growing in its native home, is said to have produced fronds eleven feet in length and three feet two inches in width. This plant had altogether fifty fronds, which covered an area of eighteen and a half feet.[1] In Tasmania this fern forms the great feature in the fern valley. Humboldt considers it singular that no mention is made of arborescent ferns in the classic authors of antiquity, the first distinct reference being by Oviedo, in the early part of the sixteenth century. However graceful and elegant some of the palms may be in their foliage and the grandeur of their crested forms, these cannot

---

[1] Lowe's "Ferns, British and Foreign," vol. viii. pl. 126.

be compared for beauty with the deeply-cut and infinitely diversified and subdivided fronds of the larger ferns. All that the palms may claim for excess in height, or bulk of trunk, over the tree ferns, is amply compensated in the latter by the beauty and grace of their crown of feathery fronds.

Seaweeds are the most gigantic of cryptogamic plants, and of these the most noteworthy is the large Macrocystis of the antarctic seas (*Macrocystes pyrifera*). D'Urville says that it grows in eight, ten, and even fifteen brasses of water, from which depth it ascends obliquely, and floats along the surface nearly as far; this gives a length of 200 feet. Dr. Hooker (now Sir Joseph) says: "In the Falkland Islands, Cape Horn, and Kerguelen's Land; where all the harbours are so belted with its masses that a boat can hardly be forced through, it generally rises from eight to twelve fathom water, and the fronds extend upwards of one hundred feet upon the surface. We seldom, however, had opportunities of measuring the largest specimens, though washed up entire on the shore; for on the outer coasts of the Falkland Islands, where the beach is lined for miles with entangled cables of *Macrocystis*, much thicker than the human body, and twined of innumerable strands of stems coiled together by the rolling action of the surf, no one succeeded in unravelling from the mass any one piece upwards of seventy or eighty feet long; as well might we attempt to ascertain the length of hemp fibre by unlaying a cable. In Kerguelen's Land the length of some pieces which grew in the middle of Christmas Harbour was estimated at more than three hundred feet.[1] He afterwards alludes to what he considered the largest specimens seen, in what is believed to be forty fathoms water, and streaming along the surface, to a probable total length of about 700 feet. The report that this seaweed sometimes attains a length of fifteen hundred feet is probably exaggerated, although it may be true that "it grows up from a depth of forty-five fathoms to the surface, at a very oblique angle, and even when of no great breadth, make excellent natural floating breakwaters."

None of the remaining cryptogamia attain to any extraordinary size. Neither floating mosses nor dendritic forms exceed two or three feet; and lichens only extend to about the same dimensions in the most exaggerated examples. Fungi have not yet produced a Titanic species, for the largest agaric yet known is inferior in expanse to a lady's parasol; and the great puff ball (*Lycoperdon giganteum*) has not yet attained the dimensions of a somnolent sheep. Amongst the lower cryptogamia we

---

[1] "Cryptogamia Antarctica," p. 158.

have many examples of the infinitely little, but not of the infinitely great.

Whether we study plant life in its largest or its most minute manifestations, in its simplest or most eccentric forms, through its normal development or exhibiting strange phenomena, we are induced to join with Horatio Smith in his exquisite hymn—

> 'Neath cloistered boughs, each floral bell that swingest
>   And rolls its perfume on the passing air,
> Makes Sabbath in the fields, and ever ringest
>   A call to prayer.
>
> Not to the domes, where crumbling arch and column
>   Attest the feebleness of mortal hand,
> But to the fane, most catholic and solemn,
>   Which God hath planned.
>
> To that cathedral, boundless as our wonder,
>   Whose quenchless lamps the sun and moon supply;
> Its choir the wind and waves, its organ thunder,
>   Its dome the sky.
>
> There, as in solitude and shade I wander,
>   Through the green aisles, or stretched upon the sod,
> Awed by the silence, reverently ponder
>   The ways of God.

CHAPTER XVII.

TEMPERATURE.

WITHOUT concerning ourselves greatly as to the general temperature of plants, we may premise that the accepted opinion is in favour of the conclusion that it is more equable than that of the surrounding air; that at night, or in winter, it is above, and in midday, or in summer, it is below the atmospheric temperature. Most of those who have made experiments have come to the conclusion that trees with thick trunks have a temperature lower than that of the air during great heat, and higher during extreme cold. Dr. Hooker made some observations in India, and was of opinion that the temperature of the fluids in a plant coincided with that of the soil at the spot whence the largest absorption was derived. That a shaddock fruit maintained the same temperature at mid-day with the atmosphere at 110°, as at midnight with the thermometer at 68°. He remarked that, "when the surface sand in the Soane Valley was heated to 110° the fresh juice of *Calotropis* plant was only 72°. This latter temperature he found at fifteen inches depth in the soil where the plant grew. The power which the plant has in maintaining a low temperature of 72°, though the main portion, which is subterranean, is surrounded by a soil heated between 90° and 100°, is remarkable, and is no doubt proximately due to the rapidity of evaporation from the foliage, and consequent activity of the circulation. Its exposed leaves maintained a temperature of 80°, nearly 25° lower than the similarly exposed sand and alluvium." The inference is, that the liquids taken up by the roots, being at the degree of heat which the soil possesses, at that depth tends to warm the tree in the cold season, and to cool it, in comparison with the air, in the warm season.

Apart from this question of general temperature we are concerned chiefly with the great increase of heat evolved by plants under certain conditions, especially at germination, and during flowering. That is, the phenomena of increased temperature under special circumstances. In animals the heat of the body is maintained by a process analogous to combustion. Oxygen combines with carbon and forms carbonic acid, which latter is thrown off, the change or oxidation being accompanied by the evolution of heat. As it is in the combustion of carbon so is it in the conversion of carbon in the animal body, and so also in plants, under special conditions, when oxidation is greatly increased heat is evolved, chemical changes take place, and the burning log, the breathing animal, and germinating plant all exhibit the same phenomenon of carbon in combustion.

A familiar example of the evolution of heat during germination is furnished in the process of "malting" the grain of barley. Growth is stimulated by moisture, and a large number of seeds being collected together it is easy to experience the increase of temperature caused during the process. The chemical change which the seeds undergo, the absorption of oxygen, the state of slow combustion, the amount of heat evolved, are all easily demonstrated. Thus we ascertain that the change is a chemical one, the starch of the seed by acquiring oxygen becomes soluble and saccharine, this kind of decomposition being accompanied by increase of temperature. The process is essential to the growth of the plant. The starch was insoluble, and therefore incapable of nourishing the young embryo. By acquiring oxygen it becomes soluble and growth proceeds, until checked artificially by drying, and the starchy "barley" is converted into the sugary "malt." That which is here effected artificially is simply the ordinary course of nature.

From this process we learn that there is a chemical change, accompanied by evolution of heat, to a greater or less extent, in all seeds during germination. So, also, at a subsequent period, namely, that of flowering, certain chemical changes take place, which are equivalent to decomposition, in which oxidation takes place, and heat is evolved during the process.

We are chiefly concerned here in the phenomenon of the evolution of heat at the time of flowering, for although, as in the case of germination, it undoubtedly takes place, more or less, in all plants, it is only under favourable conditions that the temperature is raised to an appreciable extent. The most suitable condition for observing the heat evolved during germination is when a large number of seeds are collected together; so, also, the most favourable condition for the determination of the amount of heat evolved at the period of flowering is when a large number of flowers are associated together. This will account for the high temperature determined in certain plants to be presently alluded to, the results being proportioned to the number of associated flowers.

The evolution of heat at the time of flowering has been observed most frequently, and with the greatest satisfaction, in plants of the arum family, in which a large number of flowers are collected together at the base of the spadix, and these are surrounded by and enclosed within an envelope, or spathe, which prevents the rapid dissemination of the heat engendered. This structure is sufficiently represented in our common indigenous *Arum maculatum*, called "Lords and Ladies," for illustration (fig. 83). This phenomenon was first observed by Lamarck, in 1777, but without any precise determination of the heat experienced. In 1800, Sennebrier measured with a thermometer the heat evolved in the common arum, and found it 8·6′ cent. In the early part of the same century Hubert states that a thermometer placed in the centre of five spadices of *Arum cordifolium*, in the Isle of France, stood at 131° Fahr., and in twelve at 142½° Fahr., while the temperature of the air was only 74-75°. This showed an elevation of 56° and 68°.[1] Schultz observed the flowers of *Caladium pinnatifolium*, at Berlin, in 1828, and M. Treviranus published

Fig. 83.—Wake-robin (*Arum maculatum*).

the results of investigations on several species of *Arum*, in 1829. Gœpperd, in 1832, found the temperature of the spadix of *Arum dracunculus* rose to

---

[1] Hubert, in "Bory de St. Vincent Voyage," ii., p. 68.

31°, Fahr., above the temperature of the surrounding air. In 1834 M. Brongniart observed the elevation of temperature in *Colocasia odora* as 19·8°, Fahr., above that of the conservatory in which it was growing. Van Beek and Bergsma examined the same species in 1828, and found an elevation of 50°, Fahr., above the surrounding air, by means of a thermo-electric apparatus.[1]

In 1839 Vrolik and Vriese made numerous observations, and found that the maximum of several hundreds of experiments was from 48° to 57° Fahr.[2] These bring us to the memoir of Dutrochet, in which, after recounting the labours of his predecessors, he narrates his own experiences up to 1840, chiefly on the common wild Arum,[3] in which he found an elevation of temperature of from 25° to 27° Fahr. From all these experiments, made by different individuals and in diverse ways, we ascertain that there is a great elevation of temperature in the plants of the Arum family at the period of flowering, but the precise amount varies with the observers, the highest being from 50° to 68° Fahr. for the larger species, and proportionately less for the smaller ones.[4] The greatest

---

[1] Brongniart " Nouv. Ann. du Mus.," iii., 145.
[2] "Ann. des Sci. Nat.," 2nd ser., xi., p. 65.
[3] "Ann. des Sci. Nat.," 2nd ser., xiii., pp. 5 and 65 (1840).
[4] For a summary of these observations, see Balfour's " Class Book of Botany," p. 520.

heat was obtained at or shortly after the opening of the spathe, or the climax of flowering. Temperatures ascertained at different periods of the day would necessarily be influenced more by the condition of the flowers than by the precise hour. This will account for the maximum being fixed at different hours by different observers. Subsequent experiments conducted by M. Garreau at Lille demonstrated the great consumption of oxygen which accompanied the elevation of heat and its proportionate increase. When the mean heat was seven degrees, sixteen volumes of oxygen were consumed per hour; when the mean heat rose to twelve, the consumption of oxygen increased to twenty-one volumes; and when the mean heat had attained seventeen and a half, the volumes of oxygen consumed exceeded twenty-seven. The quantity of carbonic acid evolved is in direct proportion to the oxygen absorbed, and the degree of chemical action which takes place determines the amount of heat.

We shall be prepared to concede that, after all, it is not so remarkable that here and there we meet with records of an elevation of temperature at the time of flowering, in plants where the natural conditions are favourable, as that these records are not more numerous and explicit. Mr. N. E. Brown states that on one occasion the living spadix of *Philodendron Williamsii*, which had flowered at Kew, was brought to him in a condition in which it was uncomfortably hot to the hand, but he had no ready means of ascertaining the precise temperature. Mr. Nicholson has observed *Philodendron sagittæfolium* with the anthers nearly ready to dehisce, and which exhibited a rise of temperature from 69°, that of the stove, to 81° Fahr. Also of *Philodendron eximium*, when the house was at 82°, showed an elevation of 92°. Seeing that the latter are aroids, in which numerous flowers are associated, the rise of temperature was comparatively small.

Somewhat more striking results were shown some years ago by Mr. W. H. Tillet, of Norwich, on an aroid growing in his conservatory, in which he observed a manifest increase of temperature, and found, by testing with a thermometer, that the elevation exceeded those above alluded to. It may be accepted as a general result of numerous experiments, that, in the large aroids, an increase of temperature of fully 30° Fahr. may be anticipated, which, under exceptionally favourable circumstances, may reach as much as 50°.

In palms and their allies the flowers are produced in dense masses, and these are often wholly or partially surrounded by an envelope, so that the physical conditions are very similar to those of the aroids; yet opportunities do not often arise for determining the heat evolved during the flowering. Mr. Nicholson determined recently the elevation of temperature in the ivory-palm (*Phytelephas macrocarpa*) at Kew. "On April 20th, at one p.m., the temperature of the house was 68° Fahr., the bulb of a thermometer, which had been suspended for some time near the plant in question, was placed in the centre of the cream-coloured inflorescence, and the mercury almost instantly rose to 92°, showing an increase in temperature of 24°. The following day, at the same hour, the thermometer registered 72° in the house, and, when placed in the same position in the centre of the inflorescence, only rose to the same height as that reached the preceding day, viz., 92°. As the drawn-out end of the bulb prevented it from actually touching the convex ovaries, a small incision was made in one of these, and the thermometer then rose to 94°."[1]

The same observer also tested another allied plant (*Carludovica plumieri*) and found the thermometer rise from 73° to 90°, but the plant was not in good condition, for "the long barren stamens had already changed from creamy white to cinnamon colour, and the spathe had commenced to decompose, although not three hours had elapsed since the flowers had opened."

"Development of Heat in Phytelephas," "Journ. Bot.," x. (1881), p. 154.

Dr. De Vriese has also referred to a high temperature obtained at Burtenzorg, in the male cones of *Cycas circinalis*, but does not state the precise amount; he says that the elevation always took place between six and ten p.m., and was accompanied by a strong smell.[1]

The evolution of heat in other plants, where the flowers are produced singly, and not enclosed in a spathe or envelope, is not only less, but more difficult of determination than in agglomerated flowers. Wherever a number of flowers approximate, as in composite plants, greater heat has been detected. Amongst the species which have been tested may be mentioned the flowers of a *Cistus*, in which three degrees were registered above the surrounding air. In geranium as much as six degrees are said to have been determined. Saussure found by a thermometer —scarcely a satisfactory medium—that the tuberose rose half a degree, the flowers of a gourd from 1° to 3° Fahr., and a *Bignonia* only 1°. The flowers of *Victoria regia* were tested at Hamburg when the temperature of the house was 70° 7′ Fahr., and the flowers found to be 80° 3′. On another occasion, when the air was 72° 5′, the flower had risen to 105° 1′, or the rather extraordinary increase of about 33°.[2] If this determination is an accurate one

---

[1] Hooker's "Kew Gardens Miscellany," iii., p. 186.
[2] Balfour's "Class Book of Botany," p. 519.

it becomes almost inexplicable, and should at least receive some corroboration, especially when compared with the results of an examination of the flowers of *Nymphæa stellata*, another water-lily, in which the maximum elevation was little over 1° Fahr. Of the flowering heads of composite plants we have accounts of but two, the capitulum of the cotton-thistle (*Onopordum acanthium*), in which about 1° 5′ Fahr. is recorded, and in a number of flower-buds of *Anthemis chrysoleuca*, the temperature rose to 2° 4′ Fahr. One result of the great stimulus which electrical science has recently received, it may be hoped, will be an extensive series of observations, with delicate appliances, to determine the variations of temperature at different periods in a large number of plants.

Chemical change takes place so rapidly in the fleshy fungi that we should have been quite prepared to find that under certain conditions an appreciable elevation of temperature has been ascertained. It seems to us surprising rather that so small a rise in temperature has been observed, than that such changes have been recorded. The larger species of *Lycoperdon*, when quite mature, will become sensibly warmer to the hand when they exhibit signs of decomposition. The finger thrust into a decaying cluster of *Agaricus melleus* will obtain decided evidence of increase of temperature. In these cases it will be the necessary accompaniment of decomposition.

Dutrochet examined growing fungi of five species, and found in all a slight elevation of temperature, but in none so much as one degree, and in some not one fifth of a degree. Probably the most favourable period was not selected, at least the subject requires further investigation   Dr. McNab has also recorded his observations on *Lycoperdon giganteum*, but in this instance the rise was not so much as would have been expected, although in excess of the amount determined by Dutrochet. It can hardly be supposed that so large a mass, undergoing rapid chemical change, does not exceed about one degree per cent. in rise of temperature.

CHAPTER XVIII.

LUMINOSITY.

THE phenomena of "luminosity" in plants are evidently variable in their causes, as predicated by the variability of the results. We have brought together examples of these manifestations from flowering and cryptogamic plants, associated somewhat in accordance with their apparent relationship, but without any effort at explanation. There is strong presumption that some of the supposed cases, of flashes of light from bright coloured flowers, may be explained optically. Others can be accounted for by no such hypothesis. The different facts seem to group themselves thus:—Flowers exhibiting electrical flashes of light on sultry evenings; plants becoming surrounded by the vapour of essential oil, which readily takes fire; roots, or rhizomes, which, under certain conditions are luminous; and fungi which are either luminous in their imperfect, or "mycelium" condition, or when fully matured.

The luminosity of flowers, under certain conditions,

has many times been affirmed, by different and independent observers, and yet still remains the subject of some doubt and uncertainty. The earliest instance is that of the daughter of Linnæus, who observed a "lightning-like phosphorescence" in the flowers of the nasturtium during a sultry tempestuous night. Another instance was recorded in 1843, when Mr. Dowden mentioned a luminous appearance in the double variety of the common marigold. This circumstance was noticed on the 4th August, 1842, at 8 o'clock p.m., after a week of very dry weather. Four persons observed the phenomenon. By shading off the declining day-light, a gold-coloured lambent light appeared to play from petal to petal of the flowers, so as to make a more or less interrupted corona around the disc. It seemed as if this emanation grew less vivid as the light declined ; it was not examined in darkness.[1] Dr. Edwin Lankester was strongly in favour of the verity of such exhibitions.[2] Another contributor says, "I have observed it frequently, and have looked for it on each succeeding summer on the double marigold, and more especially the hairy red poppy (*Papaver pilosum*), in my garden at Mosely, in Worcestershire."[3] Many years after, and another instance was recorded : "We witnessed

---

[1] "Proceedings of the British Association for 1843."
[2] "Gardener's Chronicle," 1843, p. 691.   [3] Ibid.

(June 10, 1858) this evening, a little before 9 o'clock, a very curious phenomenon. There are three scarlet verbenas, each about nine inches high, and about a foot apart, planted in front of the greenhouse. As I was standing a few yards from them and looking at them, my attention was arrested by faint flashes of light passing backwards and forwards from one plant to the other. I immediately called the gardener and several members of my family, who all witnessed the extraordinary sight, which lasted for about a quarter of an hour, gradually becoming fainter, till at last it ceased altogether. There was a smoky appearance after each flash, which we all particularly remarked. The ground under the plants was very dry, the air was sultry and seemed charged with electricity. The flashes had the exact appearance of summer lightning in miniature. This was the first time I had seen anything of the kind, and having never heard of such appearances, I could hardly believe my eyes. Afterwards, however, when the day had been hot and the ground was dry, the same phenomenon was constantly observed at about sunset, and equally on the scarlet geraniums and verbenas. In 1859 it was again seen. On Sunday evening, July 10th of that year, my children came running in to say that the "lightning" was again playing on the flowers. We all saw it, and again, on July 11th, I thought that the flashes of light were brighter than I had ever seen them before. The weather was very sultry."[1]

The tuberose has also the reputation of being luminous in a similar manner. It has been observed, so it is said, of a sultry evening, after thunder, to dart small sparks in abundance from such of its flowers as were fading.[2] The sunflower has also a like reputation, and so has the martagon lily and the evening primrose. Altogether a number of different plants have been seen to present a similar phenomenon, and the facts are attested by a long list of different individuals.

Two theories have been propounded with respect to this class of luminosity; one that it is an optical illusion, the other that the light is electric. For the former it is contended that bright flowers are always the subjects, and this exhibition takes place in the evening. On behalf of this view, it is quoted from Goethe: "On the 19th June, 1799, late in the evening, when the twilight was passing into a clear night, as I was walking up and down with a friend in the garden, we remarked very plainly about the flowers of the oriental poppy, which were distinguishable above everything else by their brilliant red, something like flame. We placed ourselves before the

---

[1] "Gardener's Chronicle," July 16, 1859, p. 604.
[2] "Science Gossip," 1871, p. 122.

plant and looked steadfastly at it, but could not see the flash again, till we chanced in passing and repassing to look at it obliquely, and we could then repeat the phenomenon at pleasure. It appeared to be an optical illusion, and that the apparent flash of light was merely the spectral representation of the blossoms of a blue-green." On behalf of the electrical view it is urged that the occurrences have been observed at times when the air has been dry and charged with electricity.

A second class of luminous appearances are of the type of an experience also of the daughter of Linnæus with the dittany. When the daughter of Linnæus one evening approached the flowers of *Dictamnus albus* with a light, a little flame was kindled without in any way injuring them. The experiment was afterwards frequently repeated, but it never succeeded ; and whilst some scientific men regarded the whole as a faulty observation, or simply a delusion, others endeavoured to explain it on various hypotheses. One of them especially which tried to account for the phenomenon by assuming that the plant developed hydrogen found much favour. At present, when this hypothesis has become untenable, the inflammability of the plant is mentioned more as a curiosum, and accounted for by the presence of etheric oil in the flowers. Being in the habit of visiting a garden in which strong healthy plants of

Dictamnus albus were cultivated, I often repeated the experiment, but always without success, and I already began to doubt the correctness of the observation made by the daughter of Linnæus, when, during the dry and hot summer of 1857 I repeated the experiment once more, fancying that the warm weather might possibly have exercised a more than ordinary effect upon the plant. I held a lighted match close to an open flower, but again without result; in bringing, however, the match close to some other blossoms, it approached a nearly faded one, and suddenly was seen a reddish, crackling, strongly shooting flame, which left a powerful aromatic smell, and did not injure the peduncle. Since then I have repeated the experiment during several seasons, and even during wet cold summers; it has always succeeded, thus clearly proving that it is not influenced by the state of the weather. In doing so I observed the following results, which fully explain the phenomenon. On the pedicels and peduncles are a number of minute reddish brown glands, secreting etheric oil. These glands are but little developed when the flowers begin to open, and they are fully grown shortly after the blossoms begin to fade, shrivelling up when the fruit begins to form. For this reason the experiment can succeed only at that limited period when the flowers are fading. The radius is uninjured, being too green to take fire, and because the flame runs along almost as quick as lightning, becoming extinguished at the top, and diffusing a powerful incense-like smell.[1]

Possibly some of the "burning bushes" of oriental story might have a similar explanation. Vague ideas of the existence of luminous plants in India and the neighbouring countries still float about as in the days of the old Hindoos and Greeks. One of these is that in Afghanistan, to the north of Nalwo, is a mountain called Sufed Koh, in which the natives believe gold and silver to exist, and in which, they say, in the spring is a bush which at night, from a distance, appears on fire, but on approaching it the delusion vanishes. In 1845 the natives of Simla were filled with a rumour that the mountains near Syree were illuminated nightly by some magical herb. It has been suggested that this might be a species of *Dictamnus*, which abounds near Gungotree and Jumnotree.

A third class of examples of luminosity consists of those mythic and uncertain legends of roots which can only be recorded and not explained, possibly in many cases due only to decomposition. Josephus says " There is a certain place called Baaras, which produces a root of the same name with itself; its colour is like to that of flame, and towards evening it

---

[1] Dr. Hahn in "Journal of Botany," 1863.

sends out a certain ray like lightning; it is not easily taken by such as would do it, but recedes from their hands."[1] The only virtue this root possesses is its supposed power in the expulsion of demons.

The root-stock of a plant from the Ooraghum jungles is said to possess the peculiar property of regaining its phosphorescent appearance when a dried fragment of it was submitted to moisture, "gleaming in the dark with all the vividness of the glow-worm, or the electric scolopendra, after having been moistened with a wet cloth applied to its surface for an hour or two, and did not seem to lose the property by use, becoming lustreless when dry, and lighting up again whenever moistened."[2]

This, or a similar plant has long been known to the Brahmins under the name of Jyotismati, and said to be produced by a species of *Cardiospermum*. Sanscrit authorities say that it is found in the Himalayas; and Major Madden found upon enquiry at Almora that there was a luminous plant well known there as Jyotismati or Jwalla-mat, which names imply the possession of light or fire. The Almora plant proved to be the roots of the fragrant khus-khus grass, of which only one in a hundred is said to be luminous at night in the rainy season. The roots

---

[1] "Wars of the Jews," book vii., cap. vi.
[2] "Proc. Royal Asiatic Society," April, 1845.

of other grasses are reputed to possess the same properties.

If we except the milky juice or sap of two or three species, such as *Euphorbia phosphorea*, said to be luminous, this catalogue will exhaust the principal recorded cases of luminosity in flowering plants; our last class, which consists of luminous fungi, furnishes numerous well authenticated instances, which might be placed in two classes, of which one would include mycelium, or the root-like filaments of fungi in an imperfect state, and the other perfect or complete fungi. Schoolboys nearly half a century ago had a strong belief in "touchwood" and perhaps the belief still lingers. This "touchwood," consisted of very rotten wood, usually from the heart of a tree, deeply penetrated with the mycelium of fungi, and luminous in the dark. We remember many a cherished morsel which was carried in the pocket, for nocturnal exhibition in the dormitory, until "the light of other days had faded," which followed after a few days. One of the most extraordinary manifestations of this class of fungi is recorded by the Rev. M. J. Berkeley. " A quantity of wood had been purchased in a neighbouring parish, which was dragged up a very steep hill to its destination. Amongst them was a log of larch, or spruce, it is not quite certain which, 24 feet long and a foot in diameter. Some young friends happened to pass up the hill at night, and were surprised to find the road scattered with luminous patches, which, when more closely examined, proved to be portions of bark, or little fragments of wood. Following the track they came to a blaze of white light which was perfectly surprising; on examination it appeared that the whole of the inside of the bark of the log was covered with a white byssoid mycelium of a peculiarly strong smell, but unfortunately in such a state that the perfect form could not be ascertained. This was luminous, but the light was by no means so bright as in those parts of the wood where the spawn had penetrated more deeply, and where it was so intense that the roughest treatment scarcely seemed to check it. If any attempt was made to rub off the luminous matter it only shone the more brightly, and when wrapped up in five folds of paper the light penetrated through all the folds on either side as brightly as if the specimen was exposed; when, again, the specimens were placed in the pocket, the pocket when opened was a mass of light. The luminosity had now been going on for three days. Unfortunately we did not see it ourselves till the third day, when it had, possibly from a change in the state of electricity, been somewhat impaired, but it was still most interesting, and we have merely recorded what we saw ourselves. It was almost possible to read the time on the face of a watch, even in its less luminous condition. We do not for a moment suppose that the mycelium is essentially luminous, but are rather inclined to believe that a peculiar occurrence of climatic conditions is necessary for the production of the phenomenon, which is certainly one of great rarity. Observers as we have been of fungi in their native haunts for fifty years, it has never fallen to our lot to witness a similar case before, though Professor Churchill Babington once sent us specimens of luminous wood, which had, however, lost their luminosity before they arrived. It should be observed that the parts of the wood which were most luminous were not only deeply penetrated by the more delicate parts of the mycelium, but were those which were most decomposed. It is probable, therefore, that this fact is an element in the case as well as the presence of fungoid matter."[1]

Another incomplete fungus growth is that called *Rhizomorpha subterranea*, which extends underneath the soil in long strings in the neighbourhood of old tree stumps, those of oak especially, which are becoming rotten, and upon these it is fixed by its branches. These are cylindrical, very flexible, branching and clothed with a hard bark, encrusting and fragile, at first smooth and brown, becoming later

---

[1] "Gardener's Chronicle," 1872, p. 1,258.

very rough and black. The interior tissue, at first whitish, afterwards of a more or less deep brown colour, is formed of long parallel filaments. The phenomena of luminosity in these fungi have been made the subject of investigation by M. Tułasne. "On the evening of the day when I received the specimens," he writes, "the temperature being about 22° C., all the young branches brightened with an uniform phosphoric light the whole of their length; it was the same with the surface of some of the older branches, the greater number of which were still brilliant in some parts, and only on their surface. I split and lacerated many of these twigs, but their internal substance remained dull. The next evening, on the contrary, this substance having been exposed to contact with the air, exhibited at its surface the same brightness as the bark of the branches. Prolonged friction of the luminous surfaces reduced the brightness and dried them to a certain degree, but did not leave on the fingers any phosphorescent matter."[1] And again: "By preserving these Rhizomorphæ in an adequate state of humidity, I have been able for many evenings to renew the examination of their phosphorescence; the commencement of dessication, long before they really perish, deprives

---

[1] "Tulasne sur la Phosphorescence," "Ann. des Sci. Nat." (1848), vol. ix., p. 340, &c.

them of the faculty of giving light." The luminosity of this kind of fungus is well known to miners, and Humboldt, as well as others, have written of it in glowing terms. Different names have been given to different varieties, some of which have occurred in almost all parts of the world of which the lower vegetable productions are known.

The second group of luminous fungi are those exhibited by perfect and properly-developed species. These are, for the most part, agarics with white spores growing habitually on wood; and it is a remarkable fact, that although many other kinds with coloured spores grow on wood, all the known luminous species are referred to the same sub-genus (*Pleurotus*) in which the stem is eccentric, or obsolete, and the spores white.

One of the earliest known exotic species (*Agaricus Gardneri*) was first made known by Mr. Gardner in 1840. "One dark night about the beginning of December, while passing along the streets of the Villa de Natividate, Goyaz, Brazil, I observed some boys amusing themselves with some luminous object which I at first supposed to be a kind of large fire-fly; but, on making inquiry, I found it to be a beautiful phosphorescent species of *Agaricus*, and was told that it grew abundantly in the neighbourhood on the decaying fronds of a dwarf palm. The whole plant gives out at night a bright phosphorescent light, somewhat similar to that emitted by the larger fire-flies, having a pale greenish hue. From this circumstance, and from growing on a palm, it is called by the inhabitants ' Flor de Coco.' "

Dr. Cuthbert Collingwood has given his experience of the same, or a closely-allied species, in Borneo. " The night being dark, the fungi could be very distinctly seen, though not at any great distance, shining with a soft pale greenish light. Here and there spots of much more intense light were visible, and these proved to be very young and minute specimens. The older specimens may more properly be described as possessing a greenish luminous glow like the glow of the electric discharge, which, however, was quite sufficient to define its shape, and when closely examined, the chief details of its form and appearance. The luminosity did not impart itself to the hand, and did not appear to be affected by the separation from the root on which it grew, at least not for some hours. I think it probable that the mycelium of this fungus is also luminous, for, upon turning up the ground in search of small luminous worms, minute spots of light were observed which could not be referred to any particular object, or body, when brought to the light and examined, and were probably due to some minute portions of its

---

[1] Hooker's " Journal of Botany," 1840, ii., p. 426.

mycelium."[1] Mr. Hugh Low has affirmed that "he saw the jungle all in a blaze of light, by which he could see to read, as some years ago he was riding across the island by the jungle road, and that this luminosity was produced by an agaric."

Similar experiences are furnished from Australia, where several species of luminous agarics have been found. Drummond, writing from the Swan River[2] speaks of two species growing parasitically on the stumps of trees, with nothing particular in their appearance by day, but by night emitting a most curious light, such as he had never seen described in any book. The first species was about two inches across, and was growing in clusters on the stump of a Banksia tree. "The stump was at the time surrounded by water, when I happened to be passing on a dark night, and was surprised to see what appeared to be a light in such a spot. When this fungus was laid on a newspaper it emitted by night a phosphorescent light, enabling us to read the words round it, and it continued to do so for several nights with gradually decreasing intensity as the plant dried up." Subsequently he found a second species, sixteen inches in diameter, and a foot in height, weighing about five pounds. "This specimen was

---

[1] "Journal of Linnæan Society," vol. x., p. 469.
[2] Hooker's "Journal of Botany," April, 1842.

hung up inside the chimney of our sitting-room to dry, and, on passing through the apartment in the dark, I observed the fungus giving out a most remarkable light, similar to that described above No light is so white as this, at least none that I have ever seen. The luminous property continued, though gradually diminishing, for four or five nights, when it ceased on the plant becoming dry. We called some of the natives and showed them this fungus when emitting light; the room was dark, for the fire was very low and the candles extinguished, and the poor creatures cried out 'Chinga,' their name for a spirit, and seemed afraid of it."

The agaric of the olive-tree (*Agaricus olearius*) is found in the south of Europe, and has been subjected to an exhaustive examination.[1] It is of itself very yellow, reflects a strong brilliant light, and remains endowed with this remarkable faculty whilst it grows, or at least while it appears to preserve an active life, and remains fresh. Tulasne was of opinion that it was really phosphorescent of itself, and not indebted to any foreign production for the light it emits. It is unnecessary to multiply examples, in which the phenomena are uniform in their character. There is not the slightest ground for supposing that any hallucination, or optical illusion,

---

[1] Tulasne, "Annales des Sci. Nat." (1848), ix., p. 340.

can be pleaded here, the manifestations being so decided, so numerous, so well authenticated, and so widely distributed. One of the most recent additions has been a small species from the Andaman Islands; several species have now been recorded from different parts of the Australian colonies; Gaudichaud found one in Manilla, and Rumphius another in Amboyna. Dr. Hooker believes them to exist in the Sikkim Himalayas; and we have already mentioned their occurrence in Brazil and the Indian Archipelago.

We might add to these the species of *Polyporus*, mentioned by Mr. Worthington Smith, such as *Polyporus annosus*, found in the Cardiff coal-mines, the light of which was sufficient for the men to "see their hands by," and could be detected at a distance of twenty yards. *Polyporus sulfureus*, which the same observer has seen exhibiting the phenomenon.[1] Perhaps, also, some others, of which the records are uncertain, as *Corticium cæruleum*, and the unusual circumstance of a luminous myxogaster, recorded by the Rev. M. J. Berkeley, in the "Gardener's Chronicle."

From these examples it will be clear that fungi exhibit luminous properties, both in their imperfect and perfect conditions. That the light is of that

---

[1] See also "Fungi, their Nature, Uses," &c., by M. C. Cooke, p. 105.

peculiar character which is observed in the slow combustion of phosphorus, and from this resemblance it has been termed phosphorescent. It may be that some hypercritical quibbler has disputed publicly the applicability of the term "phosphorescence" to the light emitted by fungi, on the ground that "no phosphorus has been detected." Perhaps his student-life was passed so much abroad that he has forgotten much of his mother tongue. "Phosphorescence" implies no presence of phosphorus, but simply "luminous, or shining with a faint light, unaccompanied by sensible heat," hence no apology is necessary for the use of a perfectly legitimate term with its general and acknowledged interpretation.

The phenomena of light and heat in plants have not as yet received all the investigation which the subject demands. As to the latter, it becomes a question whether the luminosity is an inherent quality of certain species, since it has only been observed in a few, or whether it is an electric condition, depending largely on the atmosphere at the time. The facts at present ascertained do not permit us to suggest any theory, all we can do is to take note of the circumstances, and trust to the future for their elucidation.

CHAPTER XIX.

MYSTIC PLANTS.

MANY plants were in former times, and especially in superstitious eras, and amongst imaginative people, invested with a mystical importance, and often held in veneration as sacred. We have preferred to class them as "mystic," though sometimes they better deserve denomination as "sacred." Some have doubted whether flowers were ever worshipped, although no one has doubted their having been regarded as symbols, and introduced as such in religious ceremonies. Of our own customs there are some which may be attributed to a similar origin. No one would dispute that the use of evergreens in church decorations were symbolic of everlasting life. That white flowers at weddings were to be held as types of purity. That the planting of the yew in churchyards had a symbolic intent. In fact, that we still have our mystic plants.

In oriental countries flowers have a deeper meaning, and a more emphatic language, than with us. Imagination may run riot in Persia and India, but the love of flowers is beautifully exemplified amongst these people. Sir George Birdwood has given an illustration when, in writing of the Victoria Garden, Bombay,[1] he says, "Presently, a true Persian, in flowing robe of blue, and on his head his sheepskin hat, 'black, glossy, curled, the fleece of kar-kul,' would saunter in, and stand and meditate over every flower he saw, and always as if half in vision. And when at last the vision was fulfilled, and the ideal flower he was seeking found, he would spread his mat and sit before it until the setting of the sun, and then pray before it, and fold up his mat again and go home. And the next night, and night after night, until that particular flower faded away, he would return to it, and bring his friends in ever-increasing troops to it, and sit and sing and play the guitar or lute before it, and they would altogether pray there, and after prayer still sit before it, sipping sherbet, and talking the most hilarious and shocking scandal, late into the moonlight; and so again and again every evening until the flower died. Sometimes, by way of a grand finale, the whole company would suddenly rise before the flower, and serenade it together, with an ode from Hafiz, and depart."

In the Hindu religion bright-coloured or fragrant flowers take a prominent place as offerings to the gods, whilst the leaves or flowers of other plants are held sacred for special reasons, either historical, or

---

[1] Sir G. C. M. Birdwood, in "Athenæum."

for their fancied resemblances to mystical objects. The Trimurti, or representative of the Trinity, has two plants dedicated to it, the bael tree (*Ægle marmelos*) and the cratæva (*Cratæva religiosa*).[1] Both these trees have trifoliate leaves, and, like the shamrock, may be held to represent the Trinity.

The national legend of Krishna is popular all over India, and a kind of basil (*Ocymum sanctum*) is sacred to him as well as to Vishnu. This is also a white-flowered aromatic plant, receiving special attention, and worshipped daily.[2] According to the story, this hero is said to have gambolled with the milkmaids of Brindabun under the kadamba tree (*Nauclea cadamba*), and the ball-shaped yellow flowers are held to be particularly sacred to him. It is held to be the holiest flower in India, and is extensively imitated in the native jewellery ornaments. The same hero is reported to have fascinated the milkmaids by playing on his celebrated flute under a bakula tree (*Mimusops elengi*), and the small yellow fragrant flowers are now dedicated to him as well as to Siva. The parejati (*Erythrina indica*) may be regarded as a mystical, though not a sacred, tree.

---

[1] See also on this subject, "The Industrial Arts of India," by Sir G. C. M. Birdwood, C.S.I. (1880), p. 85, &c.

[2] The Malays strew this plant with reverence over the graves of their dead.

This flower was supposed to bloom in the garden of Indra, in Heaven, and the two wives of Krishna are said to have quarrelled for the exclusive possession of this flower, which their husband had stolen from the celestial garden. Since it was stolen by Krishna it has been under a curse, and dwells upon the earth as one of the least of the flowers, and is never used for worship. This accounts for its absence from the long catalogue of sacred flowers.

In the Hindu mythology, Kamadeva is the god of love, the analogue of Cupid, and is represented with his bow and arrows. The myth alleges that these arrows were tipped with five flowers, all of which are therefore held sacred to this god. They are (1) the champa (*Michelia champaca*), a tulip-shaped yellow flower, with a strong aromatic smell, of the magnolia family, supposed by some to have been introduced into India from China : (2) the mango flower (*Mangifera Indica*) : (3) the bulla (*Pavonia odorata*), a sweet-scented flower of the mallow family : (4) the flower of the clearing-nut (*Strychnos potatorum*) : and (5) the nagkesur (*Mesua ferrea*), with flowers white externally, and yellow filaments inside the corolla, having an odour resembling that of the wild briar. Some other authorities exclude the clearing-nut flower, and substitute that of the bela (*Jasminum sambac*), with beautifully fragrant white flowers. The screw pine (*Pandanus odoratissimus*) is also, for some reason, sacred to Kamadeva. The pollen of the flowers is most profuse, and has a faint peculiar odour. It is collected, and sold at the bazaars, being scattered over the bride at marriage ceremonies. This custom seems partly to prevail on account of the odour, and partly on account of its mystic relationship to the god of love. Attar of Keora flowers and Keora water are favourite Indian perfumes.

The brilliant asoca (*Saraca Indica*), with its large clusters of orange-red flowers, is dedicated to Siva, to whom also other and mostly yellow flowers are offered, such as the "chandra malika" (*Chrysanthemum Indicum*), the cadamba, already alluded to, and the bakula, as well as the superb crimson flowers of the bandhuca (*Ixora bandhuca*), and the fragrant jasmines (*Jasminum sambac* and *Jasminum undulatum*), the gunda (*Gardenia florida*), oleander (*Nerium odorum*), and some others. It can be readily imagined that flowers, remarkable for their beauty, bright colouring, or fragrance, would present themselves to the minds of an oriental people as fitting tributes to be laid on the shrines of their gods. Such as do not conform to these features are usually connected in some manner with the history of the mythical being to whom they are sacred, or are supposed to retain in their flowers, fruits, or leaves, some mystical resemblance to well-known symbols of the attributes of the god to whom they are dedicated.

To avoid tedium we shall omit reference to all the remaining flowers, which are dedicated to members of the Hindu pantheon, with the exception of the water lilies, and these both in ancient India and ancient Egypt occupied a prominent place in mythology. The plants themselves were, in all probability, common to both countries nearly at the same time, and if we have come to the conclusion that the pre-eminence was given to one kind in India and to another in Egypt, this resulted probably from local circumstances and local traditions. The intimate relationship between the two has necessitated a parallel history of both, commencing with the Egyptian lotos to avoid repetition. The lotos (*Nymphæa*), writes Sir G. Wilkinson, was the favourite for wreaths and chaplets. But it is singular that, while the lotos is so often represented, no instance occurs on the monuments of the Indian lotos, or Nelumbium, though the Roman Egyptian sculptures point it out as a peculiar plant of Egypt, placing it about the figure of the god Nile; and it is stated by Latin writers to have been common in the country.[1] The distinction between these two

---

[1] Wilkinson's "Popular Account of Ancient Egyptians," vol. i., p. 56.

sacred plants will be better understood by a brief general description of both, so liable to confusion by applying the name of lotos in each instance.

The sacred lotos of the Nile figures conspicuously on the monuments, enters largely into the decoration, and seems to have been interwoven with the religious faith of the ancient Egyptians. This lotos is mentioned by the old writers as an herbaceous plant of aquatic habits, and from their combined description it is evident that some kind of water lily is intended. " When the river is full, and the plains are inundated, there grow in the water numbers of lilies which the Egyptians call lotos."[1] " The lotos so-called, grows chiefly in the plains when the country is inundated. The flower is white, the petals are narrow, as those of the lily, and numerous, as of a very double flower. When the sun sets they cover the seed-vessel, and as soon as the sun rises the flowers open, and appear above the water; and this is repeated until the seed vessel is ripe and the petals fall off. It is said that in the Euphrates both the seed-vessel and the petals sink down into the water from the evening until midnight, to a great depth, so that the hand cannot reach them ; at daybreak they emerge, and as the day comes on they rise above the water; at sunrise the flowers open, and when fully.

[1] Herodotus.

expanded they rise up still higher, and present the appearance of a very double flower."[1] "The flower is small and white like the lily, which is said to expand at sunrise, and to close at sunset. It is also said that the seed-vessel is then entirely hid in the water, and that at sunrise it emerges again."[2] "When the inundating waters of the Nile retire, it comes up with the stem like the Egyptian bean, with the petals crowded thick and close, only shorter and narrower. There is a further circumstance related concerning this plant of a very remarkable nature, that the poppy-like flowers close up with the setting sun, the petals entirely covering the seed vessel; but at sunrise they open again, and so on, till they become ripe, and the blossom, which is white, falls off."[3] "They grow in the lakes in the neighbourhood of Alexandria. I know that in that fine city they have a crown called Antinöean, made of the plant which is there named lotos, which plant grows in the lakes in the heat of summer; and there are two colours of it: one of them is the colour of a rose, of which the Antinöean crown is made, the other is called lotinos, and has a blue flower."[4] From the foregoing we arrive at the following particulars of the lotos. That it is an aquatic plant, with double poppy-like

---

[1] Theophrastus.
[2] Dioscorides.
[3] Pliny.
[4] Athenæus.

flowers, expanding in the morning and closing at night—

> Those virgin lilies all the night
> Bathing their beauties in the lake,
> That they may rise more fresh and bright
> When their beloved sun's awake.

Fig. 84.—Egyptian Lotus (*Nymphæa stellata*).

Either white, blue, or rose-coloured, for there are the latter two varieties, as expressly mentioned by one author. All these features are quite consistent with the presumption that the lotos was of a kindred to our own white water lily, which is further strengthened by what is recorded of the fruit. " The size of the seed-vessel is equal to that of the largest poppy head, and it is divided by separations in the same manner as the seed-vessel of the poppy, but the seed, which is like millet, is more condensed. The Egyptians lay these seed-vessels in heaps to perish, and when they are rotten, the mass is washed in the river, and the seed taken out and dried, and is afterwards made into loaves, baked, and used for food."[1] In the principal features, all the other authorities agree. The fruit, therefore, corresponds with that of a water lily, and, moreover, it is said to possess a farinaceous root, which was eaten. From these descriptions it is evident, as more fully discussed elsewhere,[2] that the sacred lotos of the Nile was a species of *Nymphæa*, or water lily, common in the waters of that river. When Savigny returned from Egypt after the French invasion of 1798, he brought home a *blue Nymphæa*, which corresponds closely in habit to the conventional lotos so common on Egyptian monuments.

It seems very probable that the lotos-flower, which is represented in the hands of guests at Egyptian banquets (fig. 85), and those presented as offerings to

---

[1] Theophrastus.
[2] M. C. Cooke on the " Lotus of the Ancients," in " Popular Science Review," vol. x. (1871), p. 260.

the deities, were fragrant. The manner in which they are held strengthens this probability, as there is no other reason why they should be brought into such close proximity to the nose. Savigny's blue water lily (*Nymphæa cærulea*) has just the habit and the narrow acute petals of the lotos on the monuments. The white lotos was evidently *Nymphæa lotus*, which is common to India and Egypt. Like others of its kindred, it is liable to variation, and there is a red variety, which some have called a distinct species, but Roxburgh has declared that he could see no difference between them except the colour of the flowers. The blue lotos of Savigny, which he called *Nymphæa cærulea*, seems to be the *Nymphæa stellata* of modern botanists.

Fig. 85.—Lady with lotus flower, from Theban tomb (*Wilkinson*).

Messrs. Hooker and Thomson have pronounced the opinion that "the blue water lily of the Nile and India are (like their white congener *N. lotus*) specifically the same, the most prominent difference to be found between them being the sweet scent of the African plant, and its usually more numerous petals and stamens." The fragrant blue lotos seems to be the most common one represented on the monuments, but the white one is chiefly alluded to by ancient authors.

The tamara, or lotos of India, was described by ancient authors under the name of kyamos, or Egyptian bean. These descriptions are so substantial that there is not the slightest doubt of the plant being the *Nelumbium speciosum*.[1] Nothing can be more explicit than the account given by Theophrastus. He says that "it is produced in marshes and in stagnant waters, the length of the stem, at the longest, four cubits, and the thickness of a finger, like the smooth jointless reed. The inner texture of the stem is perforated throughout like a honeycomb, and upon the top of it is a poppy-like seed-vessel, in circumference and appearance like a wasp's nest. In each of the cells there is a bean projecting a little above the surface of the seed-vessel, which usually contains about thirty of these beans or seeds. The flower is twice the size of a poppy, of the colour of a full-blown rose, and elevated above the water; about each flower are produced large leaves, of the size of a Thessalian hat, having the same kind of stem as the flower-stem. In each bean, when broken, may be seen the embryo plant, out of which the leaf grows. So much for the fruit. The root is thicker than the

---

[1] M. C. Cooke on the "Lotos of the Ancients," in "Popular Science Review," vol. x., p. 262.

thickest reed, and cellular, like the stem; and those who live about the marshes eat it as food, either raw, boiled, or roasted. These plants are produced spontaneously, but they are cultivated in beds," &c.

This plant has a sacred character amongst the Hindoos, and also in China and Ceylon. It was at one time plentiful in Egypt, whence it has now totally vanished. The representations given of it upon the monuments of ancient Egypt are far less common than those of the *Nymphæa*, equally with which it is to be found on the monuments of India. It serves for the floating shell of Vishnu and the seat of Brahma. Sir W. Jones writes of it, that "the Thibetans embellish their temples and altars with it; and a native of Nepal made prostration before it on entering my study, where the fine plant and beautiful flowers lay for examination." Thunberg affirms that the Japanese regard the plant as pleasing to the gods, the images of their idols being often represented sitting on its large leaves. In China the Shing-moo, or holy mother, is generally represented with a flower of it in her hand, and few temples are without some representation of the plant. Undoubtedly two plants are sculptured on the monuments and paintings in India, but they are easily distinguished from each other by their form. The one is a lotus, or *Nymphæa*, and the other is the *Nelumbium*. The former is dedicated to Soma, the latter to Lakshmi, the Indian

Venus, the goddess of beauty, and, as the most sacred flower, may be offered to all the gods. The conclusion to be arrived at from close investigation is, that the sacred lotos of Egypt was the *Nymphæa*, whilst the sacred lotos of India was, and still is, the *Nelumbium*. The latter was the symbol of fertility in Egypt as in India, and the god Horus, the personification of the rising sun, was decorated with a wreath of its flowers and buds, and was sometimes figuratively represented as a lotus springing from the waters.[1] There are few plants richer in association than water lilies. Their flowers are yellow in the ponds of Northern Europe, white or yellow in England, blue and fragrant in Persia and Cashmere, and red in Southern India. The Egyptian lily is white, tinted with rose, and that of India is said to have been similar, till it was stained by the blood of Siva, wounded by the Hindoo Cupid Kamadeva. It is the latter that is alluded to in Lalla Rookh:—

> As bards have seen him in their dreams
> Down the blue Ganges laughing glide
> Upon a rosy Lotos wreath.

From Egypt and India we pass to Greece and Rome, yet it is not our intention to linger here, as but little importance can be attached to the flowers of Greek and Roman mythology. They never held

---

[1] "Gardener's Chronicle," July 1, 1876, p. 7.

the same position as in the former countries, and the majority of allusions are only such as relate to the legendary origin of certain plants. This may be

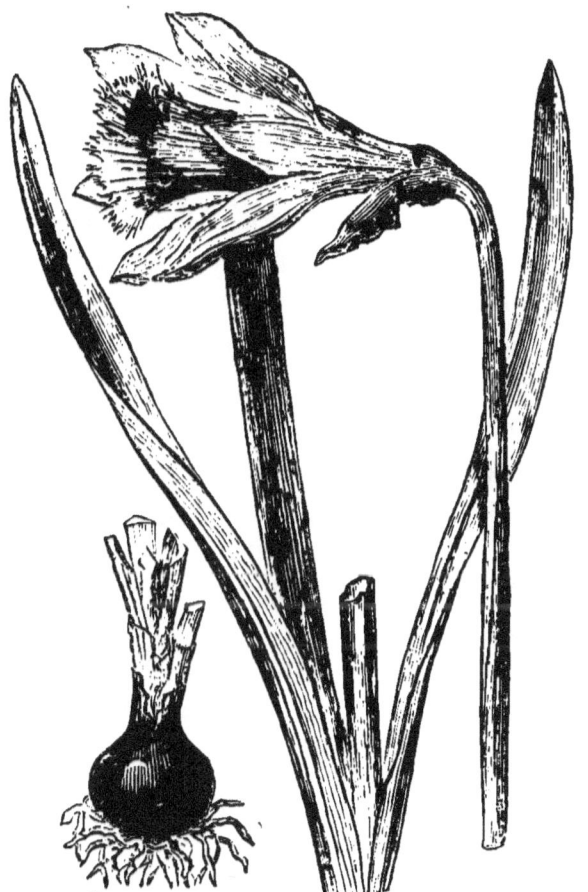

Fig. 86.—Daffodil (*Narcissus pseudonarcissus*).

illustrated by the beautiful youth Narcissus, who saw his image reflected in a fountain, and became enamoured of it, but finding that he could not reach it, grew desperate, and killed himself. His blood was changed into the flower which still bears his name. The nymphs raised a pile to burn his body, but only found a beautiful flower.[1]

Daphne fleeing from Apollo, and fearful of being caught, implored the assistance of the gods, who changed her into a laurel. Apollo crowned his head with the leaves, and for ever ordered that the tree should be sacred to his divinity. At a festival in honour of Apollo, which was held every ninth year, laurel boughs were carried in procession.

Adonis, the favourite of Venus, was fond of hunting, and in an encounter with a wild boar was so wounded that he died. The legend states that the grief of Venus was so great, that, as she wept over his dead body, the blood was transformed into roses, and the tears of the goddess herself into the anemone or "wind-flower."

> Alas the Paphian ! fair Adonis slain,
> Tears plenteous as his blood she pours amain.
> But gentle flowers are born, and bloom around
> From every drop that pours upon the ground ;
> Where streams his blood, there blushing springs the rose,
> And, where a tear has dropped, a wind-flower blows.[2]

In the sacred rites of Ceres, the Athenian matrons

---

[1] Ovid "Metamorphoses," iii., v. 346.
[2] Bion, Idyl I., 62.

strewed their couches with the leaves of the chaste tree (*Vitex agnus castus*) for the purpose of banishing impure thoughts, and hence the tree is said to have derived its name. It is added that the ancient physicians regarded the plant as an agent in securing chastity.

The dedication of the fruits of the earth to the gods in the numerous festivals, of the vine to Bacchus, and even of flowers, offer so few points of interest that we may leave their investigation to more loving hands. There is, nevertheless, an illustration of an old mythic story, which, whilst it demolishes all the poetry of the Promethean legend, exemplifies how a very simple circumstance could be transformed by the imagination into a romance.

The Ferula of the ancients was the *Ferula communis* of Crete, an umbelliferous plant, which may be compared with our wood angelica, or hog-weed Tournefort writes: " The hollow of the stem is occupied by pith, which, being well dried, takes fire like a match, without injuring the outer portion, and is therefore much used for carrying fire from place to place. Our sailors laid in a store of it. This custom is of the highest antiquity, and may explain a passage in Hesiod, where, speaking of the fire that Prometheus stole from heaven, he says that he brought it in a *Ferula*, the fact being probably that Prometheus invented the steel that strikes fire from flint, and used the pith of the *Ferula* for a match, teaching men how to preserve fire in these stalks." Alluding to this passage, Sir Wm. Hooker says— "that is, Prometheus invented the tinder-box." Unpoetical as such an explanation is, it undoubtedly comes very near the truth, and reduces a very romantic story to the poor level of an ordinary mechanical invention.

The transition from Greece and Rome to the early monkish legends associated with the Christian faith is not a very abrupt one, and if in some cases they may seem trivial, they will serve to show how minds but partially relieved from paganism exhibited a tendency to revert to the old mythical stories, and invest plain facts or simple precepts with the accessories of a pagan age. Teaching by fable or parable is a privilege which orientals have ever taken advantage of, and against it no just complaint can be made, provided that the fables are taught as fables, and not as absolute fact. This may be illustrated by a legend of the Cedar of Lebanon, which is thus recorded:—

"When Seth, the son of Adam, was sent by his dying parent to fetch the 'oil of mercy' from Paradise, he saw from the gate of that glorious garden, which an angel opened for him without permitting him to enter, a Cedar of Lebanon, with branches borne high towards Heaven. The tree seemed to typify the great disaster of Adam's early career. It stood there stricken and leafless, and yet suggesting hope—for the legend is of Christian origin—since a child in glittering raiment was seated on its top, the symbol of hope for all future generations."

This ancient legend—the dream, perhaps, of a Syrian hermit—shows that the cedar of Lebanon, the timber-tree of the temple built on Zion, was held in high estimation, and exercised the fancy. The story proceeds that Seth received from the angel three seeds of that tree which he beheld still standing upon the spot where sin had been first committed, but standing there blasted and dead. He carried the seeds home, placed them in the mouth of the dead Adam, and so buried them. And here the natural history of the legend is at fault, for the three seeds, ripened on the same tree in Paradise, produced three trees of different kinds. The truth is, the cedar of Lebanon, the cypress, and the pine, which grew from those seeds, were held in equal estimation by the recluse who dreamt this legend, and therefore the same marvellous, though inconsistent origin, was claimed for them all. Their future history is curious. Growing on the grave of Adam, in Hebron, they were afterwards most carefully protected by Abraham, Moses, and David. After their removal to Jerusalem, the Psalms were composed beneath them; and in due time, when they had grown together and united into one giant tree, they or it—for it was now one tree, a cedar of Lebanon—was felled by Solomon for the purpose of being preserved for ever as a beam in the Temple. But the design failed; the king's carpenters found themselves utterly unable to manage the mighty beam. They raised it to its intended position, and found it too long; they sawed it, and it then proved too short; they spliced it, and again found it wrong. It was evidently intended for another, perhaps a more sacred office, and they laid it aside in the Temple to bide its time. While waiting for its appointed hour, the beam was on one occasion improperly made use of by a woman named Maximella, who took the liberty of sitting on it, and presently found her garments on fire. Instantly she raised a cry, and, feeling the flames severely, she invoked the aid of Christ, and was immediately driven from the city and stoned, becoming in her death a pro-Christian martyr.

In the course of an eventful history the predestined beam became a bridge over Cedron, and, being thrown into the Pool of Bethesda, it proved the cause of its healing virtues. Finally, it became the Cross, was buried in Calvary, exhumed by the Empress Helena, chopped up by a corrupt church, and distributed.[1] Little more can be said for this than that it reads like a wild dream, and, like most dreams, with very little

---

[1] "Gardener's Chronicle," January 13, 1877.

"moral" at the end of it. Undoubtedly both Jews and Christians look upon the cedar of Lebanon with feelings very much akin to veneration, as the Hindoos look upon their own cedar, the deodar (*Cedrus deodara*), but veneration is one thing, and adoration is another, neither being improved by an admixture of superstition.

The apple has a widely extended mystical history. "The myths concerning it," as Mr. Conway has indicated, "meet us in every age and country. Aphrodite bears it in her hand as well as Eve. The serpent guards it, the dragon watches it. It is celebrated by Solomon; it is the healing fruit of Arabian tales. Ulysses longs for it in the gardens of Alcinous; Tantalus grasps vainly for it in Hades. In the prose Edda it is written that Iduna keeps in a box apples which the gods, when they feel old age approaching, have only to taste to become young again. It is in this manner that they will be kept in renovated youth until the general destruction. Azrael, the Angel of Death, accomplished his mission by holding it to his nostril; and, in the folklore, Snowdrop is tempted to her death by an apple, half of which a crone has poisoned, but recovers life when the apple falls from her lips. The golden bird seeks the golden apples in many a Norse story, and when the tree bears no more, 'Frau Bertha' reveals to her favourite that it is because a mouse gnaws at the tree's root. Indeed, the kind mother-goddess is sometimes personified as an apple-tree. But oftener the apple is the tempter in Northern mythology also, and sometimes makes the nose grow so that the pear alone can bring it again to moderate size."[1]

The association of the temptation of Eve with the apple is traditional, and not scriptural. The conception of a divinely-endowed tree guarded by a serpent makes its appearance in the myths of many ancient races.[2] In Russia the vine is sometimes represented as the Tree of Knowledge. In India it is also a climbing plant, the soma (*Sarcostemma viminale*), which is identical with the homa of the Persians. He who drinks of its juice never dies. Some authors have identified it with the "Tree of Life which grew in Paradise."

The sanctity of the oak has a remote antiquity. From the oracular oak of Dodona to the sacred oaks of the Druids it was held profoundly sacred. "The tree under which Abraham was said to have received his heavenly visitors, the "oak of mourning" under which Deborah was buried, the oak under which Jacob hid the idols at Shechem—the same probably with that near the sanctuary under which Joshua

---

[1] "Mystic Trees and Flowers" in "Fraser's Magazine," Nov., 1870, p. 590.
[2] See "Tree and Serpent Worship," by W. Ferguson, F.R.S.

set up a stone—the oak of Ophra under which the angel sat that spoke with Gideon, the oak on which Absolom hung, that under which Saul and his sons were buried—all preceded the period when Isaiah had to rebuke those who carved idols from oak, and when Ezekiel proclaimed the wrath of Jehovah against the idols standing under every thick oak."[1]

The cypress, of which idols were carved, was sacred as an evergreen. It received respect in Persia, and amongst the American Indians it is recorded that an aged cypress was held sacred and loaded with offerings. In Greece the cypresses were the daughters of Eteocles, hated by the goddesses they rivalled.

The myrtle has a sanctity that precedes that of any Christian saint. It was the emblem of Mars, and afterwards became the wreath of Aphrodite, because, after rising from the sea, she was pursued by satyrs and found refuge in a myrtle thicket. It is still sacred in the east. The Jews gather it for their feast of Tabernacles, and the Arabs say it is one of the three things that Adam brought with him out of Paradise.

The ash, in northern mythology, was the "tree of the universe." In Germany the linden, or lime, was the tree of the resurrection. The fir and the

---

[1] "Mystic Trees and Flowers," p. 592.

pine were held sacred by many races. In France, when St. Martin was permitted to destroy the temples, he was compelled to spare the holy fir groves.

The olive has become inseparably connected with one of the earliest records of the human race, and repeated references are made in the scriptures to its beauty. It probably needs an educated eye to appreciate the effect of its silver-like leaf, but it must be refreshing to ride through one of these groves when clothed with flowers, or when bowed down with fat and oily berries. Of all fruit-bearing trees the olive is the most prodigal of its flowers, but not one in a hundred comes to maturity. The tree is of slow growth, and except under peculiarly favourable circumstances, it bears no berries until the seventh year, nor is the crop worth much until the tree is ten or fifteen years old; then it is extremely profitable, and continues to yield fruit to extreme old age. There is little labour or care of any kind required, and, if long neglected, it will revive when the ground is dug or ploughed, and begin afresh to yield as before. The fruit is indispensable for the comfort, and even the existence, of the mass of the community in such places as Palestine, where the berry, pickled, forms the general relish to the dry bread. Early in the autumn the berries begin to fall. They are allowed to remain under the trees for some time, guarded by a watchman of the town. Then a proclamation is made by the governor that all who have trees should go out and pick what has fallen. Previous to this not even the owners are allowed to gather olives in the groves. The proclamation is repeated once or twice, according to the season. In November comes the final summons, when no olives are safe unless the owner looks after them, for the watchmen are removed, and the orchards become alive with men, women, and children. The shaking of the olive, which is always accompanied with much noise and merriment, is the severest operation of Syrian husbandry, particularly in the mountainous regions.[1] The olive undoubtedly stimulates in the mind of Israelite and Christian thoughts of momentous times and events; it is equally venerated by them for its history, but is so little a sacred or a mystic tree that perhaps even this passing allusion can scarcely be justified. The same may be said of flowers and plants alluded to in our Lord's teachings, or associated with His journeys. They have an interest, but not a superstitious interest, although in times past some of them have come to be regarded as mystic flowers.

As several species of true lilies and allied flowers grow in the plains around the Mount of Beatitudes,

---

[1] "Gardener's Chronicle," Sept. 18, 1875.

westward of Gennesaret, we cannot be sure what flower of deepest interest our Lord pointed to when He bade His hearers " consider the lilies of the field." Sir J. E. Smith, the great botanist, suggested that it was the amaryllis (*Sternbergia lutea*), whose golden flowers outshone " Solomon in all his glory ;" others have preferred to award the honour of having suggested the famous comparison to the " lily of Byzantium," or scarlet martagon lily, which decorates the plains of Galilee in early summer, when the Sermon on the Mount is believed to have been delivered, with floral pyramids of scarlet which are beautiful and conspicuous even at a distance."[1] It matters but little which particular flower, or whether both were alluded to, in the injunction; but it is some satisfaction to know that there are two flowers to be found at the spot, either of which would answer all the purposes of an illustration.

The monks in the middle ages were in the habit of carefully tending the lily of the valley, in the belief that it was the true " flower of the field," and it has always been in the folklore of England an emblem of purity, and connected in some way with holiness, as, for instance, in the legend of St. Leonard, who fought with a dragon for three days, and lost much blood in the encounter, and wherever the blood

---

[1] "Gardener's Chronicle," July 1, 1876, p. 7.

of the saint fell lilies of the valley sprang up, where they still grow wild in the forest of St. Leonard. The lily of the valley was introduced early into England from Southern Europe, and was largely employed in the decoration of churches in the twelfth and thirteenth centuries. When the devotion of the rosary was instituted by St. Dominic, the "Lady Chapels" erected in honour of the Virgin Mary were adorned in the season with lilies of the valley.[1]

The "Rose of Sharon" was not the rose of England, but the yellow-flowered Narcissus, common in Palestine and in the East generally, of which Mahomet said, "He that hath two cakes of bread, let him sell one of them for some flowers of Narcissus, for bread is the food of the body, but Narcissus is the food of the soul." It had been the flower-crown of the goddesses long before the period of its fame and high esteem. The Scripture "rose" is sometimes the oleander, sometimes the rhododendron.[2]

There is a curious monkish legend extant of the origin of the rose, although there is a prior one which dates from classic times. Sir John Mandeville relates that "a Christian maid of Bethlehem, blamed with wrong and slandered, and about to be martyred, prayed the Lord to spare her, and immediately red roses grew from the burning brands, and white roses

---

[1] "Gardener's Chronicle," July 1, 1876, p. 7. [2] Ibid., p. 8.

from the wood which was not on fire, and these," says Sir John, "were the first rosaries and roses, both white and red, that ever man saw," and henceforth the rose was the flower of martyrs, as well as an emblem of the Virgin. It has also been claimed for Mahomet that he created the rose.

Apropos of monkish legends, there is one, of Spanish origin, associated with a singular flower, current in Central America, of which Mr. J. K. Lord[1] has given the following graphic account. He says: "One of the most singular flowers growing in this pretty garden (of the Panama Railway Company) was an orchid (*Peristeria*), called by the natives 'Flor del Espiritu Santo,' or the 'Flower of the Holy Ghost.' The blossom, white as Parian marble, somewhat resembles the tulip in form; its perfume is not unlike that of the magnolia, but more intense. Neither its beauty nor fragrance begat for it the high reverence in which it is held, but the image of a dove placed in its centre. Gathering the freshly-opened flower, and pulling apart its alabaster petals, there sits the dove; its slender pinions droop listlessly by its side; the head inclining gently forward, as if bowed in humble submission, brings the delicate beak, just blushed with carmine, in contact with the

---

[1] J. K. Lord, Naturalist in Vancouver's Island.

snowy breast. Meekness and innocence seem embodied in this singular freak of nature; and who can marvel that crafty priests, ever watchful for any phenomenon convertible into the miraculous, should have knelt before this wondrous flower, and trained the minds of the superstitious natives to accept the title, the 'Flower of the Holy Ghost,' to gaze upon it with awe and reverence, sanctifying even the rotten wood from which it springs, and the air laden with its exquisite perfume? But it is the flower alone I fear they worship; their minds ascend not from 'nature up to nature's God;' the image only is bowed down to, not He who made it. The stalks of the plant are jointed, and attain a height of from six to seven feet, and from each joint spring two lanceolate leaves; the time of flowering is in June and July."

The " snipe orchis " will at once recur to us in this connection, as reminding us of a flying bird, represented in the centre of the flower, but, in this instance, without any mystical association (see fig. 45 ante).

We may allude, also, to the flowers which have been associated with the dead. The Greeks used amaranth, polyanthus, parsley, and myrtle to decorate tombs, and roses were prominent amongst funereal flowers. The latter also are planted on graves by the Chinese. In Upper Germany the graves are often covered with *Dianthus Carthu-*

sianorum, whilst in France the box is common in graveyards. In Switzerland and Tuscany the periwinkle (*Vinca minor*) is associated with the dead, and in many parts of Italy is called the "flower of death."

In Goethe's "Faust," Margaret plucks a flower, and picks off the petals, one by one, saying meanwhile, "He loves me, he loves me not!" This custom is a revival of an old one recorded by Theocritus, who says that the Greeks took the petal of a corn poppy, and laying it on the thumb and forefinger of one hand, slapped it with the other. If it gave a crack, it was a sign that their lovers loved them, but if it failed, they were disappointed. This was called a telephion, and a goatherd laments that he had tried whether his Amaryllis loved him, but "the telephion gave no crack."

The association of passion flowers with the passion of our Lord (as the name indicates) dates from monkish times. Dr. Masters is of opinion[1] that the species called *Passiflora incarnata* "is the one in which the semblance of the parts of the flower to the instrument of our Lord's passion was first observed. The cross, the scourge, the hammer, the nails, the crown of thorns, even ten of the apostles—Judas, who betrayed, and Peter who denied, being absent—all

---

[1] "Gardener's Chronicle," 1870, p. 1,214.

may be seen by the imaginative in these flowers. Monardes (1593) was the first to call attention to this peculiarity. Soon afterwards the plant was in cultivation at Bologna and at Rome. There is some little confusion as to the exact date, but it may safely be said to have been in cultivation in Italy before 1609. Thence it probably was introduced into Belgium, and is known to have been grown in this country in 1629. Parkinson figures it under the name of "Maracoc sive clematis virginiana—the Virginia climber." He associates it with clematis, because, as he says, "unto what other family or kindred I might better conjoin it I know not." He calls it the "surpassing delight of all flowers;" but he had very little sympathy with the imaginary description of Monardes, as will be seen from the following extract: "Some superstitious Jesuite would fain make men beleeve that in the flower of this plant are to be scene all the markes of our Saviour's passion, and therefore call it 'flos passionis,' and to that end have caused figures to be drawne and printed, with all the parts proportioned out, as thornes, nails, speare, whippe, pillar, &c., in it and all as true as the sea burnes, which you may well perceive by the true figure, taken to the life of the plant, compared with the figures set forth by the Jesuites, which I have placed here likewise for every one to see; but these bee their advantageous lies

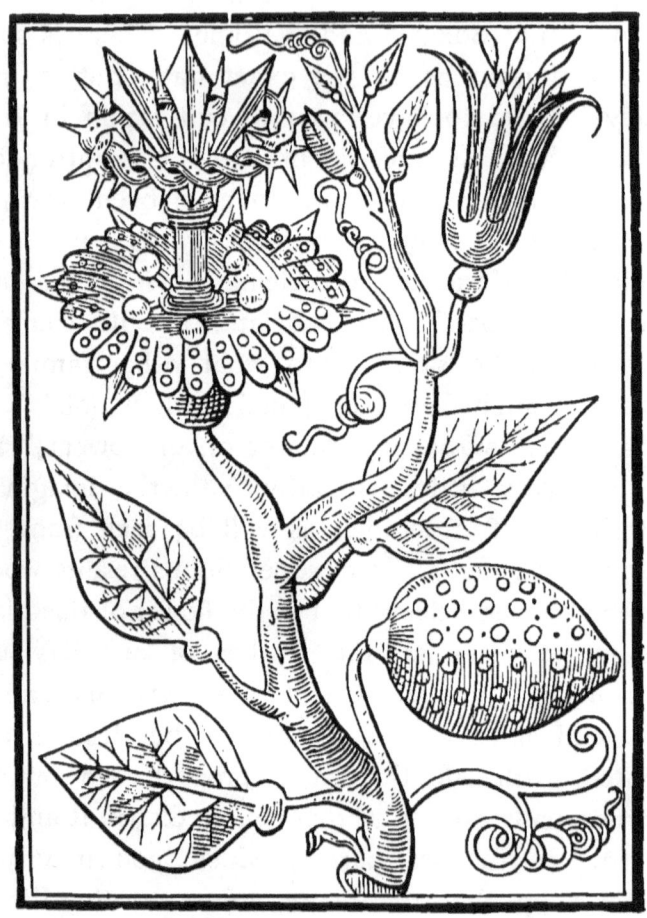

Fig. 87.—Jesuitic Maracoc, after Parkinson.

(which with them are tolerable, or rather, pious and meretorious) wherewith they used to instruct their people; but I dare say God never willed his Priests to instruct his people with lyes, for they come from the Divell the author of them. . . . . In regard whereof I could not but speake (the occasion being thus offered) against such an erroneous opinion (which even Dr. Aldine, at Rome, disproved and contraried both the said figures, and the name), and seek to disprove it, as doth (I say not almost, but I am afraid altogether) leade many to adore the very picture of such things, as are but the fictions of superstitious brains; for the flower itself is far differing from their figure, as both Aldine, in the aforesaid booke, and Robinus, at Paris, in his 'Theatrum Floræ' doe set forth; the flowers and leaves being drawne to the life, and there exhibited, which I hope may satisfie all men that will not be perpetually obstinate and contentious."

After this quotation Dr. Masters proceeds to criticise the Jesuitic figure, for he says Parkinson gives an excellent figure of *Passiflora incarnata*, " but he seems to have overlooked the fact that 'the Jesuites' figure of the Maracoc,' as copied by him, does not represent *P. incarnata* at all, but some other species, more nearly resembling *Passiflora glandulosa*, of which it has the simple leaves and the glandular footstalks. Certainly the flower in this wonderful specimen is a 'make up.' Supposing the 'corona' of threads to represent the crown of thorns, and the stamens the five nails, the Jesuit artist has just reversed their natural position; the five stamens—nails—are at the base of the column, while a terribly material crown of thorns occupies the proper place of the stamens at the top of the column. The three stigmas, too, are certainly unusually like spear-heads, so that there can be no question that Aldinus was quite correct when he stated that with the aid of a little straining of the imagination the emblems of the Passion might be as well found in a great many other flowers. It must also be remembered that no two of the older authors agree, one with the other, as to the precise significance of the several parts. By some the coronet is the type of the crown of thorns, while others see in it the 'parted vesture.' The ovary is for some the sponge dipped in gall; the stamens represent with some the nails, with others the five wounds, each author giving a slightly different version; and Ferrari compares the 'column' to the pillar to which Christ was attached, and not to the cross, because the gentle nature of the flower did not admit of its reproducing the emblem of the gibbet!"[1]

---

[1] Subsequent critical observations by Mr. A. Forsyth, in "Gardener's Chronicle" (1870), p. 1,409, do not controvert these remarks.

> I saw him as he mused one day
>   Beneath a forest bower,
> With clasp'd hands stand, and upturn'd eyes,
>   Before a *Passion flower;*
> Exclaiming with a fervent joy,
>   " I have found the Passion flower !"
>
> The passion of our blessed Lord,
>   With all his pangs and pain,
> Set forth within a beauteous flower,
>   In shape and colours plain.
>
> Up, I will forth into the world
>   And take this flower with me,
> To preach the death of Christ to all
>   As it was preached to me.

The gathering of willow catkins on Palm Sunday is the remains of a custom of the early Church in remembrance of the palm branches strewed in the way of Christ as he went up to Jerusalem. Sprigs of boxwood are still used in Catholic countries, and the willow collected on Palm Sunday is called palm by many who gather it. Why the willow should have come into use for such a purpose, has been explained in various ways. Thus, "because willow was in ancient days a badge of mourning, as may be collected from the several expressions of Virgil, where the nymphs and herdsmen are introduced sitting under a willow mourning their loves." This is hardly satisfactory, because the original palm branches were not emblems of mourning, but of triumph. A less elaborate reasoning is that "these seem to have been

Fig. 88.—Passion Flower (*Passiflora cincinnata*). "Gardener's Chronicle."

selected as substitutes for the real palm, because they are generally the only things, at this season, which can be easily procured, in which the power of vegetation can be discovered." Box was evidently in use in this country in the middle of the sixteenth century and it is possible that the use of box was discontinued on the plea that it was a Romish superstition ; although the bearing of palms was declared in 1536 "not to be contemned and cast away;" yet in Stow's Chronicle (1548) it is stated that "this yeere the ceremony of bearing palmes on Palme Sonday was left off, and not used as before." The ceremony of "blessing the box" is still continued in some of the countries of the continent.[1]
Another, and more humble plant, a kind of clover (*Medicago echinus*), found in the Levant, is held in reverence as a supposed

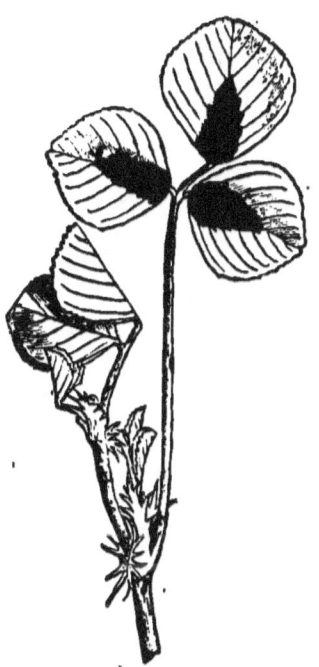

Fig. 89. — *Medicago echinus*. — "Gardener's Chronicle."

---

[1] See "Gardener's Chronicle" for a *résumé* of a sermon on one of these occasions, in which the symbolism of the box is insisted upon, April 19, 1873, p. 543.

memento of the Passion, with the symbol of the wounds on the leaves, and the crown of thorns in its spiny fruits.

Of other customs which remain as simple ceremonies, with little meaning, it may safely be predicated that they had in past times a mystic association, now forgotten. The use of holly, ivy, and mistletoe, as Christmas decorations, are of this kind, in the latter case with a date anterior to the introduction of Christianity. Whatever may be its position now, the mistletoe was in former times a mystic plant; and, as Schouw says, "It is not a matter of surprise that a plant of such peculiar aspect, and which occurs in such a remarkable position as the mistletoe, should have awakened the attention of various races, and exerted influence over their religious ideas. It played an especially important part among the Gauls. The oak was sacred with them; their priests abode in oak forests; oak boughs and oak leaves were used in every religious ceremony, and their sacrifices were made beneath an oak tree; but the mistletoe, when

Fig. 90.—Mistletoe (*Viscum album*).

it grew upon the oak, was peculiarly sacred, and regarded as a divine gift. It was gathered, with great ceremony, on the sixth day after the first new moon of the year : two white oxen, which were then for the first time placed in yoke, were brought beneath the tree ; the sacrificing priest (Druid), clothed in white garments, ascended it, and cut off the mistletoe with a golden sickle ; it was caught in a white cloth held beneath, and then distributed amongst the bystanders. The oxen were sacrificed, with prayers for the happy effects of the mistletoe. A beverage was prepared from this, and used as a remedy for all poisons and diseases, and which was supposed to favour fertility. A remnant of this seems to exist still in France ; for the peasant boys use the expression, 'au gui l'an neuf,' as a new year's greeting. It is also a custom in Britain to hang the mistletoe to the roof on Christmas eve ; the men lead the women under it, and wish a merry Christmas and a happy new year. Perhaps the mistletoe was taken as a symbol of the new year, on account of its leaves giving the bare tree the appearance of having regained its foliage."[1]

One of the strangest of mystic plants is the "Mandrake." Some belief in its power was evidently current amongst the Hebrews. Josephus gives an account of the custom in Jewish villages of pulling

---

[1] Schouw, " Earth Plants and Man," p. 218.

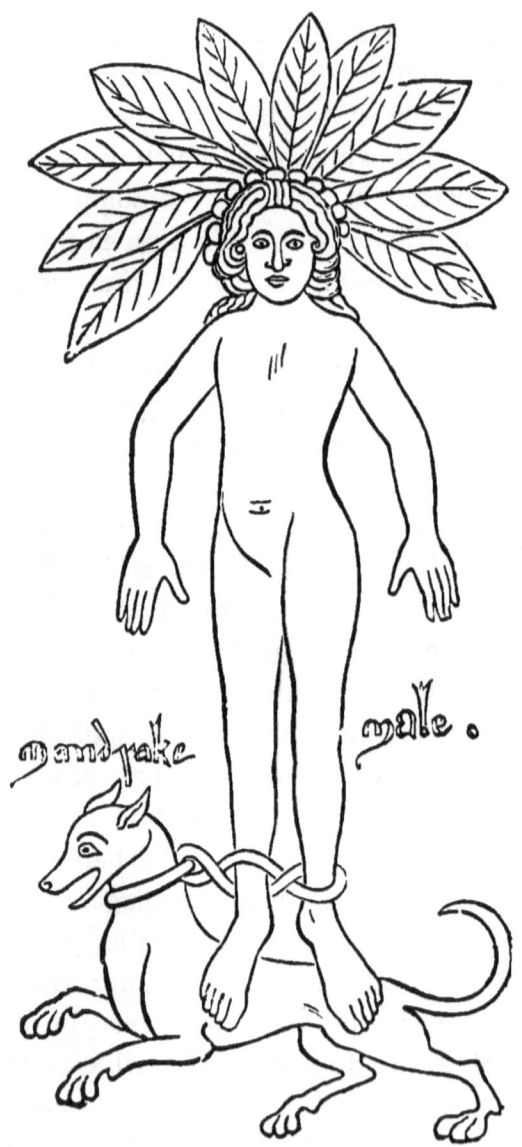

Fig. 91.—Male Mandrake.

up the root by means of a dog, which is killed by its shriek. This is the salient feature of the superstition, " To procure it, one must cut away all rootlets to the main root ; to pull up that would cause death to any creature hearing it. So one must stop his ears carefully, and, having tied a dog to the root, run away. The dog is then called, and pulling up the root, is instantly killed."

It was believed in France and Germany that the mandrake sprang up where the presence of a criminal had polluted the ground, and was sure to be found near a gallows. Having got the root, it must be bathed every Friday, kept in a white cloth in a box, and then it would procure manifold benefits. A letter, written by a burgess of Leipsic to his brother in Riga (in 1675), has been preserved, and this contains the popular notion of the time as to the virtues of the mannikin, earth-man, or mandrake. It recites that the writer had heard of his brother that in " thy home affairs hast suffered great sorrow ; that thy children, cows, swine, sheep, and horses, have all died; thy wine and beer soured in thy cellar, and thy provender destroyed ; and that thou dwellest with thy wife in great contention." He then proceeds to say that he went to those who understood such things, and they told him that these evils proceeded not from God, but from wicked people, and this was the remedy. 'If thou hast a mandrake, and bring it into thy house, thou shalt have good fortune." So he had one purchased for him for sixty-four thalers, and sent it to him as a present, with these instructions: "When

Fig. 92.—Female Mandrake.

thou hast the mandrake in thy house, let it rest three days without approaching it; then place it in warm water. With the water afterwards sprinkle the animals and sills of the house, going over all, and soon it shall go better with thee, and thou shalt come to thy own if thou serve the mandrake right. Bathe it four times every year, and as often wrap it in silk cloths and lay it among thy best things, and thou need do no more. The bath in which it has been bathed is specially good. When thou goest to law, put the mannikin under thy right arm, and thou shalt succeed, whether right or wrong."[1]

Curious old figures of the traditional mandrake are extant, of which we give copies. Stories of its potency, and of marvels associated with its possession, are numerous in Britain, France, and Germany.

> Or teach me where that wondrous mandrake grows
> Whose magic root, torn from the earth with groans
> At midnight hour, can scare the fiends away,
> And make the mind prolific in its fancies.[2]

In a French work (dated 1718) a peasant is said to have possessed a bryony root of human shape, which he received from a gipsy. He buried it at a lucky conjunction of the moon with Venus, in spring, and on a Monday, in a grave, and sprinkled it with milk in which three field-mice had been drowned. In a

---

[1] "Mystic Trees and Flowers," in "Fraser's Magazine," December, 1870.
[2] Longfellow's "Spanish Student," p. 92.

month it became more human-like than ever; then he placed it in an oven with vervain, wrapped it afterwards in a dead man's shroud; and so long as he kept it he never failed in luck at games or work. The root of the white bryony has, during later times, been designated the "mandrake," but the precise time or history of its substitution for the genuine mandrake is obscure. In different parts of Europe fragments of the old superstition still linger, and bits of the root are cherished as charms, love-tokens, as a preventive from night-mare, or a protection from bad men and evil spirits, or even for the old virtues attributed to it by the Jews.

It would not be difficult to occupy an entire chapter with allusions to flowers and plants, or some of their parts, which have had a reputation in times past of being associated with the world of spirits, as philtres or love-charms, as a protection against witchcraft, or as possessing some mysterious virtue. Such was the Saint John's Wort (*Hypericum perforatum*), gathered on the eve of St. John the Baptist Day, and hung over doors and windows as a charm against storms, thunder, and evil spirits, or carried on the person as a protection against witchcraft and enchantment, the gathering of fern-seed on Midsummer's Eve and many others, curious enough in themselves, but which have become "popular antiquities."

A somewhat kindred subject, which has never been

exhaustively treated, is the "language of flowers," in its broadest and most philosophical aspect. It is more true of such countries as Persia and India than of England and France, that every indigenous flower has become the symbol of some attribute or idea, and hence it speaks a language to the natives of those countries of which we have not learnt the alphabet. The Hindoo or the Parsee sees a symbol in every object and in every act of his life; *our* interest in flowers is more sensual; we admire their colour, their form, their odour, and, if these gratify us, we are content. Perhaps we might with profit study the language of flowers in the East, and find something to learn from the Parsee or the Hindoo.

> Bring flowers to the shrine where we kneel in prayer,
> They are nature's offering, their place is there;
> They speak of hope to the fainting heart;
> With a voice of promise they come and part;
> They sleep in dust, through the wintry hours;
> They break forth in glory—bring flowers, bright flowers!

CHAPTER XX.

FLOWERS OF HISTORY.

Some little latitude for gossip may perhaps be accorded to us for a final chapter, even if it should not concern itself much with scientific fact. Confessedly, we are proposing to enter the regions of tradition and romance, with no design of illuminating dark pages of history, or giving a new rendering to old myths. Tales of the nursery, and similar juvenile eras, are apt to cling about one, in spite of more serious studies, through many a decade. After a long journey a traveller may be permitted to describe an adventure or two, and narrate some of the legends of the country through which he has passed. It will not be wholly trivial to ascertain, if it can be done, what are the plants which as emblems or myths are associated with old stories. The rose, thistle, and shamrock may be familiar enough in name, but it will be seen that it is not quite so easy to determine which is the thistle and what is the shamrock, as might at first be imagined. Little national predilections are apt to come in the way, so that what reason might be disposed to accept, prejudice is fain to dispute.

Reasonably and loyally we commence with the rose, which old Gerarde says "doth deserve the chiefest and most principal place among all flowers whatsoever, being not only esteemed for his beautie, vertues, and his fragrant and odoriferous smell, but also because it is the honor and ornament of our English scepter, as by the conjunction appeereth in the uniting of those two most royal houses of Lancaster and York."

The emblematic rose of England is not involved in much obscurity, and the period of its first assumption seems to be contained in the following record :— "The roses of England were first publicly assumed as devices by the sons of Edward III. John of Gaunt, Duke of Lancaster, used the red rose for the badge of his family, and his brother Edward, who was created Duke of York in 1385, took a white rose for his device, which the followers of them and their heirs afterwards bore for distinction in that bloody war between the two Houses of York and Lancaster. The two families being happily united by Henry VII., the male heir of the house of Lancaster marrying Princess Elizabeth, the eldest daughter and heiress of Edward IV. of the House of York, 1486, the two roses were united in one, and became the royal badge of England."[1]

---

[1] Hugh Clark's "Introduction to Heraldry," 13th ed. (1840), p. 172.

Before the adoption of the rose, the broom was the badge of the House of Plantagenet. Tradition says that the name is derived from this circumstance, *planta* and *genista* being combined. The latter (*Genista*) was the botanical name for the "broom" before the present one (*Sarothamnus*) was adopted. The name of "Plantagenet," another account says, was first assumed by Geoffrey, Earl of Anjou, the husband of Matilda, Empress of Germany, who, having placed a sprig of the "broom" in his helmet on the day of battle, originated the surname, which was bequeathed to his descendants.

Fig. 93.—Broom (*Sarothamnus scoparius*).

The hawthorn is associated also with the Royal House of England, and was the badge of the Tudors. On the authority of Miss Strickland, this was its origin. When the body of Richard III., who was slain at Redmore Heath, was plundered of its armour and ornaments, the crown was hidden by a soldier in a hawthorn bush. It was soon found and carried back to Lord Stanley, who placed it on the head of his son-in-law, saluting him by the title of Henry VII., whilst the victorious army sang the "Te Deum." In memory of this event it is said that the House of Tudor assumed as a device a crown in a bush of fruiting hawthorn. There is an old proverb:—

Cleave to the crown, though it hang in a bush,

which appears to allude to this tradition.

Stow gives an account of King Henry VIII. and Queen Katherine riding a-Maying from Greenwich to the high ground of Shooter's Hill, accompanied by many lords and ladies, but we doubt if this had any relation to the tradition above quoted. In all the old May-day customs gathering the hawthorn had a prominent place. Brand, in his "Antiquities,"[1] gives a long account of the customs in vogue on May-day, and their supposed relationship to the ancient floralia, and subsequent association with Robin

---

[1] Brand, "Antiquities," vol. i., pp. 212 to 270.

Hood and his merry men. The first of May was also called Robin Hood's day, and even Bishop Latimer failed to secure an audience on that day, for all the parish had gone abroad to gather for Robin Hood, so that he "was fain to give place to Robin Hood and his men."

> We have been rambling all this night,
> And almost all this day;
> And now returned back again,
> We have brought you a branch of May.

The historical associations of the "forget-me-not" (or *Myosotis arvensis*) are narrated to the following effect. Miss Strickland, writing of Henry of Lancaster, says, this royal adventurer, the banished and aspiring Lancaster, appears to have been the person who gave to the "forget-me-not" its emblematical and poetical meaning, by uniting it, at the period of his exile, in his collar of SS., with the initial letter of his *mot* or watchword, "souveigne, vous de moy," thus rendering it the symbol of remembrance. Henry is said to have exchanged this token of goodwill and remembrance with his hostess, who was at that time wife of the Duke of Bretagne. If this be a true tradition, then we must bid farewell to the poetical romance of the drowned knight, who being carried by the stream, as he gathered some of these flowers for his lady, made use of the expression since associated as its name.

Many other trees and flowers have from time to time been associated, historically, with events which have transpired in this country; but Boscobel Oak and Glastonbury thorn, and such mementoes must be passed over, as our limits are reached, and we must hasten to the final page.

There has been continued controversy as to the plant with three leaflets which furnished St. Patrick with his familiar illustration of the doctrine of the Trinity. Some have affirmed that this, the Irish shamrock, is the plant we call wood-sorrel,[1] whilst others, with whom most Irishmen agree, maintain that it is the white clover.[2] The visit of the saint to the Emerald Isle is supposed to have taken place about the year 433, whereas the white clover is of comparatively recent introduction into Ireland, so that it could not have been *that* plant which apparently was so ready at hand to illustrate the saint's discourse. In Morison's history, written at the commencement of the seventeenth century, it is said that " the Irish willingly eat the herb shamrock, being of a sharp taste, which they snatch out of the ditches.[3] This description, however applicable it may be to the wood-sorrel, is not equally so to the white clover. The Irish sham-

---

[1] Oxalis acetosella.   [2] Trifolium repens.
[3] Fynis Morison's "History of the Civil Wars in Ireland, between 1599 and 1603."

rock was certainly a plant having leaves composed of three leaflets, and as a four-leaved shamrock was supposed to possess magical virtues, it may be assumed that it was not common. This would be true also of the wood-sorrel, but it is not true of the white clover, for a leaf possessed of a supplementary leaflet is by no means uncommon. In fact, if one of these two plants is to be regarded as the veritable shamrock, the evidence is very strongly in favour of the wood-sorrel, notwithstanding the national predilection for the clover.

The Scotch emblem the thistle, has been the subject of much controversy, both as to its origin and the particular species which is symbolical. The tradition has often been cited which carries its origin back to the time of the Danish invasion. "In a night assault, a bare-footed Dane trod on a thistle, and uttering a cry from the sudden pain, the sleeping Scotch were timeously aroused, and succeeded in defeating the enemy. Henceforth the thistle was elevated to its present distinction."[1] Sir Harris Nicholas traces the badge to James III., for, in an inventory of his jewels, thistles are mentioned as part of the ornaments.[2]

According to Pinkerton, the first authentic mention of the thistle as the badge of Scotland is in Dunbar's poem entitled "The Thrissell and the Rois," written

---

[1] "Notes and Queries," v., p. 281.   [2] Ibid., i., p. 90.

in 1503, on the occasion of the marriage of James IV. with Margaret Tudor. Hamilton of Bargowe expressly states that the plant was the "Monarch's choice,"[1] and Sir D. Lindsey in 1537, mentions it as the emblem of James V.

The botanical question, "which is the true Scotch thistle?" was investigated by Dr. G. Johnston,[2] and his conclusions are those now generally accepted. What is denominated by gardeners the "Scotch Thistle"[3] is an introduced plant, and not a native, and, though it has had advocates, and is planted round the grave of Burns

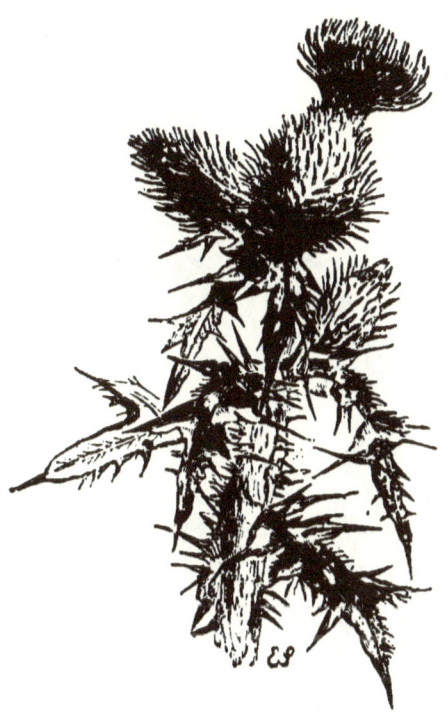

Fig. 94.—Cotton Thistle (*Onopordum acanthium*).

---

[1] Notes in Dunbar's Poems, vol. ii., p. 219.
[2] Johnston's "Botany of the Eastern Borders," p. 130.
[3] Onopordum acanthium.

in Dumfries, it could scarcely have been the traditionary thistle of Scotland. A young chieftain in the Hebrides pointed out another plant (*Carduus eriphorus*) as the Scotch thistle. At Inverness Sir James Grant said that the Scotch thistle was the only one that drooped (*Carduus nutans*); and, finally, Sir William Drummond maintained that no particular thistle, but any thistle the poet or painter chose, was the national flower of Scotland. Whether it was a thistle armed with spines or not was contested, and this induced Dr. Johnston to seek a solution by an examination of the figures impressed on the money of the kings of Scotland. "Now, the first who so marked his money was James V., and on the coins of

Fig. 95.—Musk Thistle (*Carduus nutans*).

his reign (1514 to 1542) the head or flower of a thistle only is represented. On a coin of James VI., of 1599, there are three thistles grouped and united at the base, whence two leaves spread laterally, and the stalk of the plant is spinous. On later coins, as on one of 1602, there is only a single head, while the leaves and spines are retained, and this figure is the same given on all subsequent coins, the form of the flower itself having suffered no change from its first adoption. "This evidence," says Johnston, "seems

Fig. 96.—Scotch coin of 1602.   *Fig.* 97.—Scotch coin of 1599.

to me to put *Carduus nutans* out of court, and the greater number of species, and very much to invalidate the claims of the *Onopordum*, but greatly to strengthen our belief that *Carduus marianus* was the chosen emblem of the national pride and character, although it must be admitted that the resemblance between the plant and the picture of the artist is somewhat postulatory. The bold motto, 'nemo me impune lacessit,' was the addition of

James VI., and *Carduus marianus* is almost the only species that would naturally suggest it; or that really deserves it, but I suspect that the reason for the preference of this species as the emblem was the fact of its dedication to the mother of Our Saviour, a drop of whose milk having fallen on the leaves,[1] imprinted the accident in those white veins which so remarkably distinguished them. The period at which the thistle was emblazoned was rife in those religious associations and adoptions."[2] In favour of this view an argument may be derived from the fact of the "Blessed thistle" having been cultivated in the neighbourhood of castles in Scotland, about whose ruins it is now found.[3]

The simple daisy, with all its simplicity, is very nearly a royal flower. It was once of great renown, and was called in England "Herb Margaret," or day's eye, but in France it was Marguerite, a name it still bears. It was the device of the unfortunate Margaret of Anjou, and when this queen was in

---

[1] "The purple-flowered Lady's Thistle, the leaves of which are beautifully diversified with numerous white spots like drops of milk, is vulgarly thought to have been originally marked by the falling of some drops of the Virgin Mary's milk on it, whence, no doubt, its name Lady's—*i.e.*, Our Lady's Thistle. —Brand's "Popular Antiquities," i., p. 48.

[2] Johnston's "Botany of the Eastern Borders," p. 131.

[3] Professor Balfour, "The Bass Rock" (1848), p. 419.

prosperity her nobles wore wreaths of this flower, and had it embroidered on their robes. Another Margaret, the friend of Erasmus, Margaret of Valois, had the daisy flower worn in her honour. It is said that she was called by her brother, Francis I., his "Marguerite of Marguerites."

The Lily of France, viz., the heraldic lily, is evidently one of those corruptions which are not uncommon when the origin or meaning of an emblem is forgotten or has become corrupted. It is generally considered that the Fleur-de-lys is a corruption of Fleur-de-Luce, which, again, was in itself the representative of Fleur-de-Louis. The flower itself was the common purple iris, and not a white lily, and the whole history is apparently summed up in the tradition that when Louis VII., King of France was setting out on his crusade to the Holy Land, he chose the purple iris as his heraldic emblem. Thenceforth it became the Flower of Louis, or Fleur-de-Louis, subsequently Fleur-de-Luce, and in more degenerate times it settled into Fleur-de-lys. The similarity of colouring in the purple iris of Louis and Napoleonic violet is a strange coincidence.

It has been believed that the association of the violet with the Bonaparte dynasty originated in this wise. When Napoleon I. left France for Elba it is generally understood that he said that he would return again in the violet season. During his absence, in the villages about Paris, as well as on the banks of the Lake of Geneva, the violet was the secret symbol by which the people denoted their favourite chief and recognised each other. They also wore rings of a violet colour, with the device—"It will appear again in spring" (Elle reparoitra au printemps). When asked, "Do you like the violet?" If the answer was "Oui" (yes), the inference was that the answerer was not a confederate; but if the answer was "Eh bien!" (well), they recognised a brother conspirator, and completed his sentence, "It will appear again in spring."

The friends of Bonaparte generally wore watch-ribbons, &c., of a violet colour, and he was toasted by the name of General or Corporal Violet among his adherents from the time of his quitting France until his return. When Napoleon I. re-entered the Tuileries on March 20, 1815, after his escape from Elba, his friends saluted his return with the flower of the season—violets—in token of welcome. From that time it continued the Napoleonic flower, so much so that after Waterloo, and the replacement of Louis XVIII. on the throne, violets became seditious wear—dangerous to sport in your button-hole. The white terror waged implacable war against the purple violet. The later Empire could hardly avoid reviving the traditions of the first, and with them violets.

> Farewell to thee, France! but when Liberty rallies
> Once more in thy regions, remember me then—
> The violet still grows in the depths of thy vallies,
> Though withered thy tears will unfold it again.
>
> *(Byron.)*

The pseudo-historical Lotophagi, or Lotos-eaters, when stripped of the romance which enveloped them, became resolved into very matter-of-fact vegetarians, living on the jujube. According to Homer they were

> A hospitable race;
> Not prone to ill, nor strange to foreign guest,
> They eat, they drink, and Nature gives the feast;
> The trees around them all their fruit produce;
> Lotos the name; divine nectareous juice!
> (Thence called Lotophagi) which whoso tastes,
> Insatiate riots in the sweet repasts,
> Nor other home, nor other care intends,
> But quits his house, his country, and his friends.

By comparison of the ancient authors who have mentioned the subject, we find that the Lotos was a sweet pulpy fruit of variable size, but not larger than an olive, with a hard stone (and a stoneless variety from which wine was made). There is no allusion whatever to any peculiar effects resulting from the eating of this fruit of the kind indicated by Homer, so that this portion of the story may be eliminated as poetical. Nor is there any foundation for the romance of our own Laureate of "the mild-eyed melancholy Lotos-eaters" to whom

> The gushing of the wave
> Far away did seem to mourn and rave
> On alien shores ; and if his fellow spake,
> His voice was thin, as voices from the grave ;
> And deep asleep he seemed, yet all awake,
> And music in his ears his beating heart did make.

We have discussed this subject in another place,[1] and only need to mention here the conclusion arrived at, that the Lote-bush, which gave its name to the ancient Lotophagi to this day, furnishes its fruit as food to the Arabs of Barbary, and is the *Zizyphus lotus* of botanists.

---

[1] M. C. Cooke on the "Lotos of the Ancients," in "Popular Science Review," vol. x., p. 256.

INDEX.

ABIES Nordmanniana, 159
Acacia Farnesiana, 252
Actinotus, 329
Aldrovandra vesiculosa, 67
Amorphophallus Titanum, 359
Ampelopsis hederacea, 212
Anastatica hierochuntina, 281
Apocynum androsœmifolium, 120
Arachis hypogœa, 168
Arenaria rubra, 263
Aristolochia glauca, 119
Aristolochia goldieana, 362
Aristolochia grandiflora, 361
Arum maculatum, 374
Asplenium trichomanes, 163
Avena fatua, 275
Averrhoa bilimbi, 245

BALSAM, or Impatiens, 293
Bamboos, 355
Bee orchis, 269
Bertholletia excelsa, 312
Bignonia capreolata, 208
Bignonia littoralis, 207
Bignonia Tweediana, 210
Bomarea Carderi, 197
Broom plant, 448
Byblis gigantea, 70
Byttneria aspera, 292

CALTHA dionæfolia, 328
Camrunga, *Averrhoa carambola*, 224
Carludovica Plumieri, 379
Carnivorous plants, 23
Carpels of Erodium, 280

Centaurea calcitrapa, 299
Centaurea cyanea, 297
Cephalotus follicularis, 115
Cereus giganteus, 351
Ceropegia Gardneri, 191
Ceropegia Sandersoni, 192
Change of vegetation, 112
Christian origin of rose, 427
Cleavers, *Galium aparine*, 216
Clematis flammula, 200
Clematis vitalba, 200
Cobœa scandens, 206
Colocasia esculenta, 238
Coronilla rosea, 254
Corydalis claviculata, 203
Couroupita guianensis, 313
Crambe maritima, 162
Cyclamen, 180

DARLINGTONIA, 95
Dendrobium D'Albertisii, 269
Desmodium gyrans, 223
Dictamnus albus, 387
Dielytra spectabilis, 274
Digestion by Sundews, 35
Discoid samaræ, 333
Dispersion, 291
Dispersion by Birds, 315
Drosophyllum Lusitanicum, 68

ECCENTRICITIES of flowers, 267
Echinocystis lobata, 207
Entada scandens, 354
Eucalyptus globulus, 8
Eucalyptus or gum trees, 348
Euphorbia resembling cactus, 322

FERULA of the ancients, 417
Ficus repens, 217
Floating imitators, 325
Floral clock, 260
Flowers of History, 446
Flower of the Holy Ghost, 428
Flowers of willow-herb, 234
Forget-me-not, 450
Fumaria officinalis, 203
Funaria hygrometrica, 288

GAHNIA xanthophylla, 306
Giants, 345
Gyration of plants, 150

HAMAMELIS virginica, 294
Hand-plant of Mexico, 273
Harpagophytum leptocarpum, 302
Heat and germination, 372
Heat in fungi, 381
Heliotropes or sunflowers, 170
Helleborus niger, 144
Humulus lupulus, 187
Hygienic plantations, 8
Hygroscopism, 275

IMITATING samaræ, 332
Indian fig or banyan, 350
Introduction, 1
Iresine herbstii, 229
Irish shamrock, 451

LARGE leaved palms, 356
Legend of the cedar, 419
Lilium giganteum, 364
Lily of France, 457
Locomotive Loranthus, 319
Lotos of the Nile, 407
Lotos-eaters, 459
Luminosity, 383
Luminous agarics, 395
Luminous mycelium, 391
Lupinus luteus, 250
Lycopodium and Azorella, 327

MACROCYSTIS pyrifera, 368
Mahogany tree, 292
Mandrake, 439
Marigold and luminous flowers, 384
Martynia diandra, 302
Masdevallia, 271
Medicago echinus, 437
Megaclinium bufo, 236
Mesembryanthemum tripolium, 282
Meteoric flowers, 259
Mimicry, 321
Mimicry in fungi, 339
Mimosa pudica, 221
Mirabilis jalapa, 265
Mistletoe, 438
Momordica elaterium, 294
Monkey-pots, 311
Mystic plants, 401

NAPOLEONIC violet, 457
Narcissus, 415
Nelumbium speciosum, 308
Nepenthes ampullacea, 106
,, bicalcarata, 105
,, Chelsoni, 110
,, distillatoria, 109
,, Rafflesiana, 106

ŒNOTHERA biennis, 265
Oncidium zebrinum, 269
Oxalis acetosella, 183, 243
Oxalis sensitiva, 225

PACHYSTOMA Thomsoni, 269
Parachute of Tragopogon, 296
Parnassia palustris, 233
Passiflora edulis, 215
,, gracilis, 207
Passion flower, 430
Pedalium murex, 299
Pentaclethra macrophylla, 285
Phalaris canariensis, 176
Phaseolus vulgaris, 251
Philodendron, 377
Phytelephas macrocarpa, 379

Pinguicula Lusitanica, 123
Pinguicula vulgaris, 123
Pitcher plants, 99
Plants of races, 17
Polar-plant or compass-weed, 170
Polygonum convolvulus, 195
Porlieria hygrometrica, 285
Proboscidea Jussieui, 302

RAFFLESIA Arnoldi, 359
Rain tree, 13
Relative sizes, 6
Rhizomorpha subterranea, 393
Robinia pseudacacia, 182
Rock rose, *Helianthemum*, 330
Roridula dentata, 70
Rosa setigera, 216
Rose of Sharon, 427
Royal Hawthorn, 449

SACRED flowers in India, 402
Sanctity of the oak, 422
Sandbox, *Hura crepitans*, 284
Sarracenia flava, 74
,,   purpurea, 77
,,   variolaris, 72
Scarlet pimpernel, 264
Schnella excisa, 353
Scotch thistle, 452
Sea and forest, 2
Seed of Calosanthes Indica, 335
Seeds of *Mesua ferrea*, 331
Seed of Zanonia macrocarpa, 334
Selaginella lepidophylla, 287
Sensitive plants, 220
Sequoia gigantea, 346

Silene noctiflora, 266
Similar crested seeds, 336
Sleep of plants, 239
Snake nut of Demerara, 342
Snipe orchis, 269
Solanum dulcamara, 191
Species of plants, 3
Sphærobolus stellatus, 295
Stellaria media, 249
Stipa pennata, 277
,,   spartea, 279
Sundews, 23
Sunflowers, 10

TEMPERATURE, 371
The olive, 424
Thomasia solanacea, 325
Trapa bicornis, 305
,,   bispinosa, 305
Tree ferns, 367
Tribulus terrestris, 299
Trifolium repens, 246
,,   subterraneum, 165
Tropæolum, 249
Twiners and climbers, 184

UTRICULARIA clandestina, 134
,,   montana, 143
,,   neglecta, 133
,,   vulgaris, 132

VENUS fly-trap, 50
Victoria regia, 364

WELWITSCHIA mirabilis, 349

XANTHIUM spinosum, 301
,,   strumarium, 301

WYMAN AND SONS, PRINTERS,
GREAT QUEEN STREET, LINCOLN'S INN FIELDS,
LONDON, W.C.

PUBLICATIONS OF THE
Society for Promoting Christian Knowledge.

NATURAL HISTORY RAMBLES.
Intended to cover the Natural History of the British Isles in a manner suited to the requirements of visitors to the regions named.

Fcap. 8vo., with numerous Woodcuts, cloth boards, 2s. 6d. each.

IN SEARCH OF MINERALS. By the late D. T. ANSTED, M.A., F.R.S.
LAKES AND RIVERS. By C. O. GROOM NAPIER, F.G.S.
LANE AND FIELD. By the Rev. J. G. WOOD, M.A.
MOUNTAIN AND MOOR. By J. E. TAYLOR, Esq., F.L.S.
PONDS AND DITCHES. By M. C. COOKE, M A., LL.D.
SEA-SHORE (The). By Professor P. MARTIN DUNCAN.
UNDERGROUND. By J. E. TAYLOR, Esq., F.L.S.
WOODLANDS (The). By M. C. COOKE, M.A., LL.D.

MANUALS OF ELEMENTARY SCIENCE.
A Set of Elementary Manuals on the principal Branches of Science.

Fcap. 8vo., limp cloth, 1s. each.

ELECTRICITY. By Professor FLEEMING JENKIN.
PHYSIOLOGY. By F. LE GROS CLARK, Esq.
GEOLOGY. By the Rev. T. G. BONNEY, M.A., F.G.S.
CHEMISTRY. By A. J. BERNAYS, Ph.D., F.C.S.
CRYSTALLOGRAPHY. By HENRY PALIN GURNEY, M.A.
ASTRONOMY. By W. H. M. CHRISTIE, M.A., F.R.S.
BOTANY. By Professor BENTLEY.
ZOOLOGY. By ALFRED NEWTON, M.A., F.R.S.
MATTER AND MOTION. By J. CLERK MAXWELL, M A., &c.
SPECTROSCOPE AND ITS WORK (The). By RICHARD A. PROCTOR, Esq.

EARLY BRITAIN.
This is one of a Series of Books which has for its aim the presentation of Early Britain at great historic periods.

Fcap. 8vo., with Map, cloth boards, 2s. 6d.
ANGLO-SAXON BRITAIN. By GRANT ALLEN, Esq., B.A.

THE FATHERS FOR ENGLISH READERS.

A Series of Monographs on the Chief Fathers of the Church, the Fathers selected being centres of influence at important periods of Church History and in important spheres of action.

Fcap. 8vo., cloth boards, 2s. each.

LEO THE GREAT. By the Rev. CHARLES GORE, M.A.
GREGORY THE GREAT. By the Rev. J. BARMBY, B.D.
SAINT AMBROSE: his Life, Times, and Teaching. By the Rev. ROBINSON THORNTON, D.D.
SAINT AUGUSTINE. By the Rev. E. L. CUTTS, B.A.
SAINT BASIL THE GREAT. By the Rev. RICHARD T. SMITH, B.D.
SAINT JEROME. By the Rev. EDWARD L. CUTTS, B.A.
SAINT JOHN OF DAMASCUS. By the Rev. J. H. LUPTON, M.A.
THE APOSTOLIC FATHERS. By the Rev. Canon HOLLAND.
THE DEFENDERS OF THE FAITH; or, The Christian Apologists of the Second and Third Centuries. By the Rev. F. WATSON, M.A.
THE VENERABLE BEDE. By the Rev. G. F. BROWNE.

CONVERSION OF THE WEST.

A Series of Volumes showing how the Conversion of the Chief Races of the West was brought about, and their condition before this occurred.

Fcap. 8vo., cloth boards, 2s. each.

THE CELTS. By the Rev. G. F. MACLEAR, D.D. With Two Maps.
THE ENGLISH. By the above Author. With Two Maps.
THE NORTHMEN. By the above Author. With Map.
THE SLAVS. By the above Author. With Map.
THE CONTINENTAL TEUTONS. By the Very Rev. CHARLES MERIVALE D.D., D.C.L., Dean of Ely. With Map.

NON-CHRISTIAN RELIGIOUS SYSTEMS.

A series of Manuals which furnish in a brief and popular form an accurate account of the great Non-Christian Religious Systems of the world.

Fcap. 8vo., cloth boards, 2s. 6d. each.

BUDDHISM: Being a Sketch of the Life and Teachings of Gautama, the Buddha. By T. W. RHYS DAVIDS. With Map.
CONFUCIANISM AND TAOUISM. By ROBERT K. DOUGLAS, of the British Museum. With Map.
HINDUISM. By Professor MONIER WILLIAMS. With Map.
ISLAM AND ITS FOUNDER. By J. W. H. STOBART. With Map.
THE CORÂN: Its Composition and Teaching, and the Testimony it bears to the Holy Scriptures. By Sir WILLIAM MUIR, K.C.S.I., LL.D.

Society for Promoting Christian Knowledge. 3

DIOCESAN HISTORIES.

This Series, which will embrace, when completed, every Diocese in England and Wales, will furnish, it is expected, a perfect library of English Ecclesiastical History.

CANTERBURY. By the Rev. R. C. JENKINS. With Map. Fcap. 8vo., cloth boards, 3s. 6d.

CHICHESTER. By the Rev. W. R. W. STEPHENS. With Map and Plan. Fcap. 8vo., cloth boards, 2s. 6d.

DURHAM. By the Rev. J. L. LOW. With Map and Plan. Fcap. 8vo., cloth boards, 2s. 6d.

OXFORD. By the Rev. E. MARSHALL, M.A. With Map. Fcap 8vo., cloth boards, 2s. 6d.

PETERBOROUGH. By the Rev. G. A. POOLE, M.A. With Map. Fcap. 8vo., cloth boards, 2s. 6d.

SALISBURY. By the Rev. W. H. JONES. With Map and Plan. Fcap. 8vo., cloth boards, 2s. 6d.

YORK. By the Rev. Canon ORNSBY, M.A. With Map. Fcap. 8vo., cloth boards, 2s. 6d.

CHIEF ANCIENT PHILOSOPHIES.

A Series of Books which deals with the Chief Systems of Ancient Thought, not merely as dry matters of History, but as having a bearing on Modern Speculation.

Fcap. 8vo., Satteen cloth boards, 2s. 6d. each.

EPICUREANISM. By WILLIAM WALLACE, M.A.
STOICISM. By the Rev. W. W. CAPES.

COMMENTARY ON THE BIBLE.

With Maps and Plans. Crown 8vo., cloth boards, red edges, 4s.; half-calf, 10s.; whole calf, 12s. per vol.

OLD TESTAMENT.

Vol. I., containing the Pentateuch. By Various Authors.

Vol. II., containing the Historical Books, Joshua to Esther. By Various Authors.

Vol. III., containing the Poetical Books, Job to Song of Solomon. By Various Authors.

Vol. IV., containing the Prophetical Books, Isaiah to Malachi. By Various Authors.

Vol. V., containing the Apocryphal Books. By Various Authors.

NEW TESTAMENT.

Vol. I., containing the Four Gospels. By the Right Rev. W. WALSHAM HOW, Bishop of Bedford.

Vol. II., containing the Acts, Epistles, and Revelation. By Various Authors.

THE HEATHEN WORLD AND ST. PAUL.

This Series is intended to throw light upon the writings and labours of the Apostle of the Gentiles.

Fcap. 8vo., cloth boards, 2s. each.

ST. PAUL IN GREECE. By the Rev. G. S. DAVIES, M.A., Charterhouse, Godalming. With Map.

ST. PAUL IN DAMASCUS AND ARABIA. By the Rev. GEORGE RAWLINSON, M.A., Canon of Canterbury. With Map.

ST. PAUL AT ROME. By the Very Rev. CHARLES MERIVALE, D.D., D.C.L., Dean of Ely. With Map.

ST. PAUL IN ASIA MINOR AND AT THE SYRIAN ANTIOCH. By the Very Rev. E. H. PLUMPTRE, D.D., Dean of Wells. With Map.

ANCIENT HISTORY FROM THE MONUMENTS.

This Series of Books is chiefly intended to illustrate the Sacred Scriptures by the results of recent Monumental Researches in the East.

Fcap. 8vo., cloth boards, 2s. each.

SINAI: from the Fourth Egyptian Dynasty to the Present Day. By HENRY S. PALMER, Major R.E., F.R.A.S. With Map.

BABYLONIA (The History of). By the late GEORGE SMITH, Esq., of the British Museum. Edited by the Rev. A. H. SAYCE, Assistant Professor of Comparative Philology, Oxford.

GREEK CITIES AND ISLANDS OF ASIA MINOR. By W. S. W. VAUX, M.A., F.R.S.

ASSYRIA, from the Earliest Times to the Fall of Nineveh. By the late GEORGE SMITH, Esq.

EGYPT, from the Earliest Times to B.C. 300. By S. BIRCH, LL.D., &c.

PERSIA, from the Earliest Period to the Arab Conquest. By W. S. W. VAUX, M.A., F.R.S.

MANUALS OF HEALTH.

A Set of Manuals for Household Use.

Fcap. 8vo., limp cloth, price 1s. each.

ON PERSONAL CARE OF HEALTH. By the late E. A. PARKES, M.D., F.R.S.

FOOD. By ALBERT J. BERNAYS, Professor of Chemistry at St. Thomas's Hospital.

WATER, AIR, AND DISINFECTANTS. By W. NOEL HARTLEY, Esq., King's College.

HEALTH AND OCCUPATION. By B. W. RICHARDSON, Esq., F.R.S., M.D.

THE HABITATION IN RELATION TO HEALTH. By F. S. B. FRANÇOIS DE CHAUMONT, M.D., F.R.S.

THE HOME LIBRARY.

A Series of Books illustrative of Church History, &c., specially, but not exclusively, adapted for Sunday Reading.

Crown 8vo., cloth boards, 3s. 6d. each.

BLACK AND WHITE. MISSION STORIES. By H. FORDE.

CHARLEMAGNE. By the Rev. E. L. CUTTS, B.A. With Map.

CHURCH IN ROMAN GAUL (THE). By the Rev. R. TRAVERS SMITH, B.D. With Map.

CONSTANTINE THE GREAT: The Union of Church and State. By the Rev. E. L. CUTTS, B.A.

GREAT ENGLISH CHURCHMEN; or, Famous Names in English Church History and Literature. By W. H. DAVENPORT ADAMS.

JOHN HUS. The commencement of Resistance to Papal Authority on the part of the Inferior Clergy. By the Rev. A. H. WRATISLAW, M.A.

JUDÆA AND HER RULERS, from Nebuchadnezzar to Vespasian. By M. BRAMSTON. With Map.

MILITARY RELIGIOUS ORDERS OF THE MIDDLE AGES; the Hospitallers, the Templars, the Teutonic Knights, and others. By the Rev. F. C. WOODHOUSE, M.A.

MITSLAV; or, the Conversion of Pomerania. By the late Right Rev. R. MILMAN, D.D. With Map.

NARCISSUS: A Tale of Early Christian Times. By the Rev. Canon BOYD CARPENTER, M.A.

SKETCHES OF THE WOMEN OF CHRISTENDOM. Dedicated to the women of India. By the author of "The Chronicles of the Schönberg-Cotta Family."

THE CHURCHMAN'S LIFE OF WESLEY. By R. DENNY URLIN, Esq.

THE HOUSE OF GOD THE HOME OF MAN. By the Rev. Canon JELF.

THE INNER LIFE, as Revealed in the Correspondence of Celebrated Christians. Edited by the late Rev. T. ERSKINE.

THE LIFE OF THE SOUL IN THE WORLD; Its Nature, Needs, Dangers, Sorrows, Aids, and Joys. By the Rev. F. C. WOODHOUSE, M.A.

THE NORTH AFRICAN CHURCH. By the Rev. J. LLOYD, M.A. With Map.

EARLY CHRONICLERS OF EUROPE.

The object of this Series is to bring readers face to face with the sources of Early European History.

Crown 8vo., cloth boards, 4s. each.

ENGLAND. By JAMES GAIRDNER, author of "The Life and Reign of Richard III.," &c.

FRANCE. By GUSTAVE MASSON, B.A., Univ. Gallic., Assistant Master and Librarian of Harrow School, &c.

CHINA. By Professor ROBERT K. DOUGLAS, of the British Museum. With Map and Eight full-page Illustrations, and several Vignettes. Post 8vo., cloth boards, 5s.

FREAKS AND MARVELS OF PLANT LIFE; or, Curiosities of Vegetation. By M. C. COOKE, M.A., LL.D. With numerous Woodcuts. Post 8vo., cloth boards, 6s.

RUSSIA, PAST AND PRESENT. Adapted from the German of Lankenau and Oelnitz. By Mrs. CHESTER. With Map and three full-page Woodcuts and Vignettes. Post 8vo., cloth boards, 5s.

WRECKED LIVES; or, Men who have Failed. First and Second Series. By W. H. DAVENPORT ADAMS. Crown 8vo., cloth boards, 3s. 6d. each series.

SOME HEROES OF TRAVEL; or, Chapters from the History of Geographical Discovery and Enterprise. Compiled and re-written by W. H. DAVENPORT ADAMS. With Map. Crown 8vo., satteen cloth boards, 5s.

BIBLE ATLAS (The), of Maps and Plans to Illustrate the Geography of the Old and New Testaments, and the Apocrypha. With Explanatory Notes by the late Rev. SAMUEL CLARKE; also a Complete Index of the Geographical Names in the English Bible, by Mr. GEORGE GROVE. Royal 4to., cloth, 14s.

LAND OF ISRAEL (The): a Journal of Travels in Palestine, undertaken with special reference to its Physical Character. By the Rev. Canon TRISTRAM, author of "Scenes in the East," &c. FOURTH EDITION. Revised. With Two Maps, Four full-page Coloured Plates, Eight full-page Illustrations, and numerous other Engravings. Large post 8vo., cloth boards, 10s. 6d.

NATURAL HISTORY OF THE BIBLE. By the Rev. Canon TRISTRAM. Post 8vo., with numerous Illustrations, cloth boards, red edges, price 7s. 6d.

BIBLE PLACES; or, The Topography of the Holy Land. Eighth Thousand. New and Revised Edition, embracing all the recent important results of the work carried on by the Palestine Exploration Fund. By the Rev. Canon TRISTRAM. With Map. Post 8vo., cloth boards, 4s.

SCENES IN THE EAST. Consisting of Twelve Coloured Photographic Views of Places mentioned in the Bible, beautifully executed. With descriptive Letterpress by the Rev. Canon TRISTRAM, author of "The Land of Israel," &c. 4to., cloth, bevelled boards, gilt edges, 7s. 6d.

CHRISTIANS UNDER THE CRESCENT IN ASIA. By the Rev. EDWARD L. CUTTS, B.A., author of "Turning Points of Church History," &c. With numerous Illustrations. Post 8vo., cloth boards, 5s.

NARRATIVE OF A MODERN PILGRIMAGE THROUGH PALESTINE ON HORSEBACK AND WITH TENTS. By the Rev. ALFRED CHARLES SMITH, M.A., Rector of Yatesbury, Wilts. With numerous Woodcuts, and Four Illustrations printed in Colours. Crown 8vo., cloth boards, 5s.

SINAI AND JERUSALEM; or, Scenes from Bible Lands. Consisting of Coloured Photographic Views of Places mentioned in the Bible, including a Panoramic View of Jerusalem. With Descriptive Letterpress by the Rev. F. W. HOLLAND, M.A. Cloth, bevelled boards, gilt edges, 7s. 6d.

Society for Promoting Christian Knowledge. 7

ART TEACHING OF THE PRIMITIVE CHURCH (The), with an Index of Subjects Historical and Emblematic. By the Rev. R. ST. JOHN TYRWHITT. Post 8vo., cloth boards, 7s. 6d.

JEWISH NATION (A History of the), from the Earliest Times to the Present Day. By E. H. PALMER, Esq., M.A., Professor of Arabic in the University of Cambridge. With Map of Palestine, and numerous Illustrations. Crown 8vo., cloth boards, 5s.

TURNING POINTS OF GENERAL CHURCH HISTORY. By the Rev. EDWARD L. CUTTS, B.A., Author of "Constantine the Great," &c. Crown 8vo., cloth boards, 5s.

TURNING POINTS OF ENGLISH CHURCH HISTORY. By the Rev. EDWARD L. CUTTS, B.A., Author of "Some Chief Truths of Religion," "St. Cedd's Cross," &c. Crown 8vo., cloth boards, 3s. 6d.

STUDIES AMONG THE PAINTERS. By J. BEAVINGTON ATKINSON, Esq. With Seventeen full-page Illustrations on toned paper. Small 4to., cloth boards, 7s. 6d.

UNDER HIS BANNER. Being Papers on the Missionary Work of Modern Times. By the Rev. H. W. TUCKER, M.A., Secretary to the Society for the Propagation of the Gospel in Foreign Parts. Crown 8vo., with Map of the World. Cloth boards, 5s.

BEAUTY IN COMMON THINGS. Illustrated by Twelve Drawings from Nature, printed in colours. With Descriptive Letterpress. Handsomely bound in cloth, bevelled boards, full gilt side, gilt edges, 10s. 6d.

WILD FLOWERS. By ANNE PRATT. In 2 vols., containing 192 Plates, printed in colours. 16mo., cloth boards, 12s.

FLOWERS OF THE FIELD. By the late Rev. C. A. JOHNS, B.A., F.L.S. Fcap. 8vo., with numerous woodcuts, cloth boards, 5s.

FOREST TREES. By the late Rev. C. A. JOHNS. With numerous Woodcuts. New and revised edition. Post 8vo., cloth boards, 5s.

NATURAL HISTORY OF BRITISH FISHES: Their Structure, Economic Uses, and Capture by Net and Rod. By the late FRANK BUCKLAND. With numerous Woodcuts. Crown 8vo., cloth boards, 5s.

BRITISH BIRDS IN THEIR HAUNTS. Being a Popular Account of the Birds which have been observed in the British Isles, their Haunts and Habits. By the late Rev. C. A. JOHNS, B.A., F.L.S. Post 8vo., cloth boards, 7s. 6d.

OUR NATIVE SONGSTERS. By ANNE PRATT. With 72 Coloured Plates. 16mo., cloth boards, 6s.

ANIMAL CREATION (THE): A Popular Introduction to Zoology. By the late THOMAS RYMER JONES, Esq. With nearly 500 Engravings. Post 8vo., cloth boards, 7s. 6d.

OCEAN (THE). By P. H. GOSSE, F.R.S. Post 8vo., with fifty-one Illustrations, cloth boards, 4s. 6d.

Society for Promoting Christian Knowledge.

HEROES OF THE ARCTIC AND THEIR ADVENTURES (THE). By FREDERICK WHYMPER, Esq. With Map, Eight Page Woodcuts, and numerous smaller Engravings. Crown 8vo., cloth boards, 3s. 6d.

FROZEN ASIA: a Sketch of Modern Siberia. Together with an account of the Native Tribes inhabiting that Region. By C. H. EDEN, Esq., F.R.G.S. With Map. Crown 8vo., cloth boards, 5s.

AUSTRALIA'S HEROES. Being a slight sketch of the most prominent amongst the band of gallant Men who devoted their lives and energies to the cause of Science and the development of the fifth Continent. By C. H. EDEN, Esq., F.R.G.S. With Map. Crown 8vo., cloth boards, 5s.

FIFTH CONTINENT (THE), WITH THE ADJACENT ISLANDS. Being an account of Australia, Tasmania, and New Guinea, with Statistical Information to the latest date. By C. H. EDEN, author of "Australia's Heroes," &c. With Map. Crown 8vo., cloth boards, 5s.

HISTORY OF INDIA, from the Earliest Times to the Present Day. By L. J. TROTTER, Author of "Studies in Biography," &c. Post 8vo., with a Map and twenty-three Engravings, cloth boards, 10s. 6d.

ASTRONOMY WITHOUT MATHEMATICS. By Sir EDMUND BECKETT, Bart., LL.D. Post 8vo., cloth boards, 4s.

ON THE ORIGIN OF THE LAWS OF NATURE. By Sir EDMUND BECKETT, Bart., LL.D. Post 8vo., cloth boards, 1s. 6d.

CHEMISTRY OF CREATION (The): Being a Sketch of the Chief Chemical and Physical Phenomena of Earth, Air, and Ocean. By ROBERT ELLIS, M.R.C.S. A new edition, revised by Professor BERNAYS, M.D., F.C.S., &c. With numerous Illustrations. Fcap. 8vo., cloth boards, 4s.

EVENINGS AT THE MICROSCOPE; or, Researches among the Minuter Organs and Forms of Animal Life. By P. H. GOSSE, Esq., F.R.S. A new edition, revised and annotated. Post 8vo., cloth boards, 4s.

DEPOSITORIES—
LONDON: NORTHUMBERLAND AVENUE, CHARING CROSS, W.C.; 43, QUEEN VICTORIA STREET, E.C.; 48, PICCADILLY, W.; AND 135, NORTH STREET, BRIGHTON.

www.ingramcontent.com/pod-product-compliance
Lightning Source LLC
Chambersburg PA
CBHW051852300426
44117CB00006B/369